THE GEOMETRY CENTER

Mathematical Sciences Research Ins
Publications

D. Drasin   C.J. Earle   F.W. Gehring
I. Kra   A. Marden
Editors

Holomorphic Functions and Moduli I

Proceedings of a Workshop held
March 13-19, 1986

With 22 Illustrations

Springer-Verlag
New York Berlin Heidelberg London Paris Tokyo

D. Drasin
Department of Mathematics
Purdue University
West Layfayete, IN 47909
USA

C.J. Earle
Department of Mathematics
Cornell University
Ithaca, NY 14853
USA

F.W. Gehring
Department of Mathematics
University of Michigan
Ann Arbor, MI 48109
USA

I. Kra
Department of Mathematics
State University of
   New York at Stony Brook
Stony Brook, NY 11794
USA

A. Marden
Department of Mathematics
University of Minnesota
Minneapolis, MN 55455
USA

Mathematical Sciences Research Institute
1000 Centennial Drive
Berkeley, CA 94720
USA

The Mathematical Sciences Research Institute wishes to acknowledge support by the National Science Foundation.

Mathematics Subject Classification (1980): 30XX

Library of Congress Cataloging-in-Publication Data
Holomorphic functions and moduli / D. Drasin...[et al.], editors.
    p.   cm. -- (Mathematical Sciences Research Institute
    publications ; 10-<11   >)
    Includes bibliographies.
    ISBN 0-387-96766-4 (v. 1). ISBN 0-387-96786-9 (v. 2)
    1. Holomorphic functions. 2. Teichmuller spaces.
3. Quasiconformal mappings. 4. Riemann surfaces. I. Drasin, D.
(David)   II. Series: Mathematical Sciences Research Institute
publications ; 10, etc.
QA331.H66   1988   88-12381
515.9'8--dc19

Camera-ready text prepared by the Mathematical Sciences Research Institute using PC T_EX.
Printed and bound by R.R. Donnelley & Sons, Harrisonburg, Virginia.
Printed in the United States of America.

9 8 7 6 5 4 3 2 1

ISBN 0-387-96766-4 Springer-Verlag New York Berlin Heidelberg
ISBN 3-540-96766-4 Springer-Verlag Berlin Heidelberg New York

PREFACE

The Spring 1986 Program in Geometric Function Theory (GFT) at the Mathematical Sciences Research Institute (MSRI) brought together mathematicians interested in Teichmüller theory, quasiconformal mappings, Kleinian groups, univalent functions and value distribution. It included a large and stimulating Workshop, preceded by a mini-conference on String Theory attended by both mathematicians and physicists. These activities produced interesting results and fruitful interactions among the participants. These volumes represent only a portion of the papers that will eventually result from ideas developed in the offices and corridors of MSRI's elegant home.

The Editors solicited contributions from all participants in the Program—whether or not they gave a talk at the Workshop. Papers were also submitted by mathematicians invited but unable to attend. All manuscripts were refereed. The articles included here cover a broad spectrum, representative of the activities during the semester. We have made an attempt to group them by subject, for the reader's convenience.

The Editors take pleasure in thanking all participants, authors and referees for their work in producing these volumes.

We are also grateful to the Scientific Advisory Council of MSRI for supporting the Program in GFT. Finally thanks are due to the National Science Foundation and those Universities (including Cornell, Michigan, Minnesota, Rutgers Newark, SUNY Stony Brook) who gave released time to faculty members to participate for extended periods in this program.

The staff at MSRI, the beautiful surroundings and fine weather were instrumental in significantly increasing the usual pleasures of doing mathematics. We look forward to another program at MSRI.

D. Drasin
C.J. Earle
F.W. Gehring
I. Kra
A. Marden

QUASICONFORMAL MAPPINGS

Holomorphic Functions and Moduli

TABLE OF CONTENTS – VOLUME 2

P.L. Sipe
FAMILIES OF COMPACT RIEMANN SURFACES WHICH
DO NOT ADMIT n^{th} ROOTS

The nonconjugacy of certain exponential functions

BY Adrien Douady and Lisa R. Goldberg

The exponential family $E_\lambda(z) = \lambda \exp z$ where λ is a nonzero complex number has been studied extensively (see [D], [DK], [DG], [DGH]). It is known that either

— The stable set Ω_λ of E_λ consists of a single periodic cycle of stable regions and their preimages [GK], or

— $\Omega_\lambda = \emptyset$.

The latter case occurs for example when $\lambda = 1$ [M]; we remark that the proof of this extends essentially without change to all E_λ's for which λ is real and $> 1/e$. In this note we restrict our attention to these values of λ, and show:

THEOREM. *If $\beta, \gamma \in (1/e, \infty)$ and $\beta \neq \gamma$, E_β is not topologically conjugate to E_γ.*

REMARK: The weaker fact that E_β is not quasiconformally conjugate to E_γ was previously known. For the existence of such a conjugacy implies the existence of an open set U containing β and γ which parameterizes a subfamily of E_λ's with empty stable sets. This contradicts a result of Devaney ([D]).

Given $M > 0$, $S(M)$ will denote the half strip $\{z \mid \operatorname{Re} z \geq M, |\operatorname{Im} z| \leq \pi/2\}$.

LEMMA 1. *For all $\lambda \in (1/e, \infty)$ and $0 < a < 1$, there exists $M = M(a, \lambda)$ such that if $z \in S(M)$ and $|\operatorname{Im} E_\lambda(z)| \leq \pi/2$ then:*

$$Re(E_\lambda(z)) > E_{a\lambda}(\operatorname{Re} z) > 10 \operatorname{Re} z$$

PROOF: Fix M large enough so that:

(1) $\cos(\pi/\lambda \exp M) > a$
(2) $a\lambda \exp x > 10x, \quad$ for all $x \geq M$.

Research of the second author partially supported by grants from the National Science Foundation and PSC–CUNY.

2

Then:

$$\pi/2 \geq |\mathrm{Im}(E_\lambda(z))| = \lambda \exp(\mathrm{Re}\ z)|\sin(\mathrm{Im}\ z)|$$
$$\geq \lambda/2 \exp(\mathrm{Re}\ z)|\mathrm{Im}\ z|.$$

Consequently

$$\pi/\lambda \exp M \geq \pi/\lambda \exp(\mathrm{Re}\ z) \geq |\mathrm{Im}\ z|$$

and

$$a < \cos(\pi/\lambda \exp M) \leq \cos(\mathrm{Im}\ z).$$

Therefore,

$$\mathrm{Re}\ E_\lambda(z) = \lambda \exp(\mathrm{Re}\ z) \cos(\mathrm{Im}\ z)$$
$$> a\lambda \exp(\mathrm{Re}\ z)$$
$$= E_{a\lambda}(\mathrm{Re}\ z)$$
$$> 10\ \mathrm{Re}\ z.$$

$\square$

COROLLARY 2. *If* $M = M(1/2, \lambda)$ *and* $E_\lambda^k(z) \in S(M)$ *for all* k, *then* $z \in \mathbb{R}^+$.

PROOF: Lemma 1 and induction imply that $\mathrm{Re}\ E_\lambda^{k+1}(z) \geq 10\ \mathrm{Re}\ E_\lambda^k(z)$ for all k. Hence,

$$(1) \qquad\qquad \lim_{k \to \infty}\ \mathrm{Re}\ E_\lambda^k(z) \to \infty.$$

But

$$|\mathrm{Im}\ E_\lambda^{k+1}(z)| = \lambda \exp \mathrm{Re}\ E_\lambda^k(z)|\sin\ \mathrm{Im}\ E_\lambda^k(z)|$$
$$\geq \lambda \exp \mathrm{Re}\ E_\lambda^k(z)|\mathrm{Im}\ E_\lambda^k(z)|$$
$$\geq C_k|\mathrm{Im}\ E_\lambda^k(z)|$$

where C_k tends to infinity by (1). Hence, $\lim_{k \to \infty} |\mathrm{Im}\ E_\lambda^k(z)| \to \infty$, which is a contradiction. $\square$

PROOF OF THEOREM: If the theorem is false, there is a homeomorphism $\varphi : \mathbb{C} \to \mathbb{C}$ such that

$$\varphi \circ E_\beta = E_\gamma \circ \varphi.$$

LEMMA 3. *φ restricts to a homeomorphism of the positive real axis.*

PROOF: 0 is the unique omitted value for both E_β and E_γ, hence $\varphi(0) = 0$.

Furthermore, $\mathbb{R}^+$ is topologically distinguished as the only E_β-forward invariant curve which starts at 0 and tends to infinity. To see this, note that any such curve C must lie eventually in every right half plane, or it will return infinitely often to some fixed disk centered at 0 (contradicting the fact that φ is proper). Similarly, C must lie eventually in the half strip $S(M)$ for $M = M(1/2, \beta)$ (or it will return infinitely often to the left half plane). Forward invariance now implies that any z in the unbounded component of $C \cap S(M)$ satisfies $E_\beta^k(z) \in S(M)$ for all k so $z \in \mathbb{R}^+$ by Corollary 2.

Since $\mathbb{R}^+$ is analogously distinguished by E_γ, we conclude $\varphi(\mathbb{R}^+) = \mathbb{R}^+$.

$\square$

COROLLARY 4. *φ restricts to an orientation preserving homeomorphism of the negative real axis $\mathbb{R}^-$, and the lines $L_j = \{z \mid \mathrm{Im}\ z = (2j+1)\pi\}$ for $j \in \mathbb{Z}$.*

PROOF: Among preimages of $\mathbb{R}^+$, $\mathbb{R}^-$ is distinguished for both E_β and E_γ by the properties that it is distinct from $\mathbb{R}^+$ and its boundary is 0. Hence $\varphi(\mathbb{R}^-) = \mathbb{R}^-$.

Furthermore, $E_\lambda^{-1}(\mathbb{R}^-)$ consists of the lines L_j for $\lambda = \beta, \gamma$, hence the L_j's are permuted by φ. Since each L_j disconnects $\mathbb{C}$, there exists $m \in \mathbb{Z}$ such that $\varphi(L_j) = L_{j+m}$ for all j, now $\varphi(\mathbb{R}) = \mathbb{R}$ implies $m = 0$. $\square$

For all $j \in \mathbb{Z}$, denote by Σ_j the horizontal strip $\{z \mid (2j-1)\pi < \mathrm{Im}\ z < (2j+1)\pi\}$ bounded by L_j and L_{j-1}.

Define inverses to E_β and E_γ on the complement of the nonpositive real axis in $\mathbb{C}$: for $\lambda = \beta, \gamma$, $\log_\lambda^j$ will be the branch of the inverse to E_λ taking values in Σ_j.

LEMMA 5. *$\varphi \circ \log_\beta^j = \log_\gamma^j \circ \varphi$ for all j.*

PROOF: Since φ preserves each component of $\partial \Sigma_j$ in an orientation preserving manner, it preserves Σ_j as well, Lemma 5 follows immediately.

$\square$

For $p, q \in \mathbb{Z}$ define $z_\lambda^{p,q} = (\log_\lambda^0)^p(2\pi qi)$. Under iteration by E_λ, $z_\lambda^{p,q}$ spends $p - 1$ iterates in Σ_0 and then lands on the y axis at $2\pi qi$.

Note that:

$$\varphi(\beta) = \varphi \circ E_\beta(0) = E_\gamma(0) = \gamma$$

and $\log^q_\lambda(\lambda) = 2\pi q i$ for $\lambda = \beta, \gamma$. Therefore Lemma 4 implies:

COROLLARY 6. $\varphi(z^{p,q}_\beta) = z^{p,q}_\gamma$ for all p, q. $\qquad\square$

LEMMA 7. $\mathbb{R}$ is the closure of $\{z^{p,q}_\lambda\}$ for $\lambda = \beta, \gamma$.

PROOF: It suffices to prove Lemma 7 for $x > M$ where M is any constant. For convenience we take $M = 2$ so that if Rc $z > M$,

$$|(\log^j_\lambda)'(z)| \leq 1/2.$$

Fix $x > M$, a large number p, and $\lambda = \beta$ or γ. We will construct a sequence $z^{p,q(p)}_\lambda$ for $p \in \mathbb{Z}$ converging to x.

Let $q = q(p)$ be the smallest integer such that $\log 2\pi q - \log \lambda \geq E^{p-1}_\lambda(x)$. If p is large enough,

$$\log 2\pi q - \log \lambda - E^{p-1}_\lambda(x) < \pi/2.$$

Then $\log^0_\lambda 2\pi q i = \log 2\pi q - \log \lambda + 2\pi i = z^{1,q}_\lambda$ and the length of the line segment α connecting $E^{p-1}_\lambda(x)$ and $z^{1,q}_\lambda$ is less than $\sqrt{2}\pi/2$.

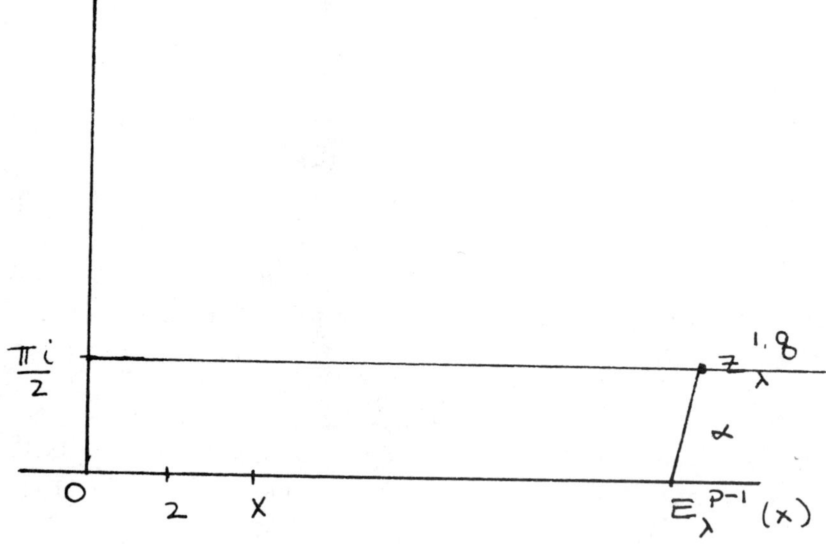

Figure 1

An inductive argument shows that if p is sufficiently large, $(\log_\lambda^0)^j(\alpha)$ is contained in the half plane Re $z > 2$ for $j = 0, \dots, p-1$, hence:

$$|x - z_\lambda^{p,q}| = |(\log_\lambda^0)^{p-1} \circ E_\lambda^{p-1}(x) - (\log_\lambda^0)^{p-1}(z_\lambda^{1,q}))|$$
$$\leq \sup_{y \in \alpha} |((\log_\lambda^0)^{p-1})'(y)||E_\lambda^{p-1}(x) - z_\lambda^{p,q}|$$
$$\leq (1/2)^{p-1}\sqrt{2}\pi/2.$$

Therefore we find some $z_\lambda^{p,q(p)}$ in every $(1/2)^{p-1}(\sqrt{2}\pi/2)$ neighborhood of x. $\qquad\qquad\square$

END OF PROOF OF THEOREM: Lemma 7 allows to choose a sequence $z_\beta^{p,q(p)}$ tending to β, and from Corollary 6 we know:

$$z_\gamma^{p,q(p)} = \varphi(z_\beta^{p,q(p)}) \to \varphi(\beta) = \gamma.$$

Assume $\beta < \gamma$ and fix $a < 1$ so that $\beta < a\gamma$. There exists, from Lemma 1, a number $M = \max(M(a,\gamma), M(a,\beta))$ such that if:

(1) $\qquad\qquad z_\beta, z_\gamma \in S(M)$ and $|\mathrm{Im}\, E_\lambda(z_\lambda)| \leq \pi/2$ for $\lambda = \beta, \gamma$

then:

(2) $\qquad\qquad \mathrm{Re}\, E_\gamma(z_\gamma) > E_{a\gamma}(\mathrm{Re}\, z_\gamma)$ and $\mathrm{Re}\, E_\beta(z_\beta) > 10\, \mathrm{Re}\, z_\beta.$

If (1) holds and Re $z_\gamma >$ Re z_β, (2) implies:

$$\mathrm{Re}\, E_\gamma(z_\gamma) > E_{a\gamma}(\mathrm{Re}\, z_\gamma)$$
(3)
$$> E_\beta(\mathrm{Re}\, z_\beta)$$
$$> \mathrm{Re}\, E_\beta(z_\beta)$$
$$> 10\, \mathrm{Re}\, z_\beta > M$$

and we conclude from (3) that $E_\lambda(z_\lambda) \in S(M)$ for $\lambda = \beta, \gamma$.

Now $E_\gamma^n(\gamma) > E_\beta^n(\beta)$ for all n and we can find a smallest N such that:

$$E_\gamma^N(\gamma) > E_\beta^N(\beta) > M.$$

If p is sufficiently large then $E_\lambda^N(z_\lambda^{p,q(p)})$ will be as close as we like to $E_\lambda^N(\lambda)$ for $\lambda = \beta, \gamma$. In particular we can assume:

$$\mathrm{Re}\, E_\gamma^N(z_\gamma^{p,q(p)}) > \mathrm{Re}\, E_\beta^N(z_\beta^{p,q(p)}) > M.$$

Applying (3) $(p - N - 1)$ times, it follows that:

$$\operatorname{Re} E_\gamma^{p-1}\big(z_\gamma^{p,q(p)}\big) > \operatorname{Re} E_\beta^{p-1}\big(z_\beta^{p,q(p)}\big).$$

But since $E_\lambda^p\big(z_\lambda^{p,q(p)}\big) = 2\pi q i$ for $\lambda = \beta, \gamma$,

$$\begin{aligned}
2\pi q &= \big|E_\gamma\big(E_\gamma^{p-1}\big(z_\gamma^{p,q(p)}\big)\big)\big| \\
&= \gamma \exp \operatorname{Re} E_\gamma^{p-1}\big(z_\gamma^{p,q(p)}\big) \\
&> \beta \exp \operatorname{Re} E_\beta^{p-1}\big(z_\beta^{p,q(p)}\big) \\
&= 2\pi q.
\end{aligned}$$

which is absurd. $\qquad\square$

REMARK: One can use symbolic dynamics to label the periodic points for E_β and E_γ; for both maps they are indexed by the set of periodic sequences of nonzero integers [**DK**]. It is natural to try to define a "conjugacy", between E_β and E_γ restricted to their periodic points, by insisting that it induce the identity map on the labelling. In view of our theorem such a map can have no continuous extension to the plane.

We close by mentioning an open question:

PROBLEM: Is there an open subset $U \subset \mathbb{C}$ parametrizing a subfamily of E_λ's whose Julia sets are equal to the entire Riemann sphere?

Adrien Douady, Ecole Normale Superieure
and
Lisa R. Goldberg, Brooklyn College, CUNY

REFERENCES

[**D**] Devaney, R., *The structural instability of* exp(z), Proc. AMS **94** (1985), 545–548.

[**DG**] Devaney, R. and Goldberg, L.R., *Uniformization of attracting basins for exponential maps*, Duke Journal (to appear).

[**DGH**] Devaney, R., Goldberg, L.R. and Hubbard, J.H., *A dynamical approximation to the exponential map by polynomials*, to appear.

[**DK**] Devaney, R. and Krych, M., *Dynamics of* exp(z), Ergodic Theory and Dynamical Systems **4** (1984), 35–52.

[**GK**] Goldberg, L.R. and Keen, L., *A Finiteness Theorem for a Dynamical Class of Entire Functions*, Ergodic Theory and Dynamical Systems **6** (1986), 183–192.

[**M**] Misiurewicz, M., *On iterates of e^z*, Ergodic Theory and Dynamical Systems **1** (1981), 103–106.

Dynamics of holomorphic self-maps of $\mathbb{C}^*$

BY LINDA KEEN

Abstract. In this paper we classify the stable components of holomorphic self-maps of $\mathbb{C}^*$ which have finitely many singular values. We use this to study a one and a two parameter family of such functions. We examine the dynamic dependence of these functions on the parameters and study the parameter spaces themselves.

1. Introduction.

In 1953 Radström [**R**] showed that among planar domains with holomorphic self-maps $f: D \to D$, the only dynamically interesting cases are: $D = \hat{\mathbb{C}}$, f rational, $D = \mathbb{C}$, f entire or $D = \mathbb{C}^* = \mathbb{C} - \{0\}$. In all other cases every point of D is a stable point for f. The dynamics of rational maps was studied by Fatou and Julia in the early part of the century and has been studied more recently using the theory of quasi-conformal mappings. In [**S1**], Sullivan applied these methods to prove that the stable components of $\hat{\mathbb{C}}$ fall into finitely many different periodic orbits, and in [**GK**], we extend these results to finite type entire functions, that is functions which fail to be a regular covering of their image at only a finite number of points. In trying to understand more generally what classes of functions admit only eventually periodic stable phenomena, we study a class of examples (see section 6) which lead us to consider a special class of functions, namely, holomorphic self-maps of $\mathbb{C}^*$. In this paper, we classify the stable components of these functions topologically and prove a finiteness theorem for this class of functions. We then show how it can be applied to study special families of entire functions which are not of finite type.

The same finiteness theorem was proved independently by Kotus ([**Ko**]). We wish to thank Bob Devaney and Lisa Goldberg for helpful conversations.

2. Preliminaries.

2.1 Classical Theory.

In what follows f is always assumed to be a holomorphic self-map of $\mathbb{C}^*$. f^n denotes the nth iterate of f. z is a *stable* point of f if there exists a neighborhood U, $z \in U$, such that $F = \{f^n|U\}$ is a normal family. The

---

This work was supported in part by NSF Grant 8503015 and PSC-CUNY Research Award, 6-65259.

stable set $Q(f)$ is the set of stable points of f. Its complement, $J(f)$ is called the *Julia* set of f. As in the rational function theory, $Q(f)$ is open, completely invariant and possibly empty, $\overline{J}(f) = J(f) \cup \{0, \infty\}$ is closed, $J(f)$ is completely invariant and never empty.

A *periodic point* of f is a point z for which there exists an integer n such that $f^n(z) = z$. The *eigenvalue* $\lambda(z)$ is defined by $\lambda(z) = (f^n)'(z)$. By the chain rule all the points in the periodic cycle, $z_k = f(z_{k-1})$, $z_0 = z$, $k = 1, \ldots, n$ have the same eigenvalue. The cycle is called *repelling* if $|\lambda(z)| > 1$, *neutral* if $|\lambda(z)| = 1$, *attracting* if $0 < |\lambda(z)| < 1$ and *super-attracting* if $|\lambda(z)| = 0$.

Fatou [**F**] proved that the Julia set of a rational function of degree greater than one or of an entire transcendental function is infinite and perfect, and that for such a rational function it is the closure of its repelling periodic points. Baker [**Ba1**] proved the same theorem for general transcendental entire functions and Battacharvya [**Bat**] proved that $\overline{J}(f)$ is the closure of the repelling fixed points in our case.

In what follows, unless specifically stated otherwise, an entire function will always mean an entire transcendental function.

LEMMA 2.1. *Let $\alpha \in J(f)$ and let U be a neighborhood of α. If K is any compact set in $\mathbb{C}^*$, there is an integer k such that all the iterates $f^n(U) \supset K$ for all $n > k$.*

Consider the commutative diagram

$$
\begin{array}{ccc}
\mathbb{C} & \xrightarrow{\ E\ } & \mathbb{C} \\
{\scriptstyle e^w}\big\downarrow & & \big\uparrow{\scriptstyle e^w} \\
\mathbb{C}^* & \xrightarrow{\ f\ } & \mathbb{C}^*
\end{array}
$$

(**)

Here we regard $\mathbb{C}$ as a covering space of $\mathbb{C}^*$. If f is given, then E is an entire map defined by the diagram. That is, $E(w) = \log(f(e^w))$, and is uniquely determined by f up to a choice of a branch of the logarithm. Any function $f(z)$ mapping $\mathbb{C}^*$ to itself has the form:

$$f(z) = z^n \exp(g(z) + h(1/z))$$

where $g(z)$ and $h(z)$ are entire functions which fix zero, n is an integer. The function $E(w)$ has the form:

$$E(w) = nw + g(e^w) + h(e^{-w}).$$

It satisfies the functional equation $E(w + 2\pi i) = E(w) + 2\pi i n$.

The iteration theories of E and f are closely related. We will use this relationship often in what follows.

LEMMA 2.2. *If z is stable for f then all its preimages are stable for E. Conversely, if w is a stable point of E, then $z = e^w$ is stable for f.*

PROOF: Suppose $z \in Q(f)$ and U is a neighborhood of z containing only stable points. Let w be any value of $\log z$ and let U be a neighborhood of w such that $\exp(U) \subset V$. Assume that $w \in J(E)$. Let R be any compact set in $\mathbb{C}$. By the statement of lemma 2.1 for entire functions, there is a k such that for $n > k$, $E^n(U) \supset R$. Hence $f^n(V) \supset \exp(E^n(U)) \supset \exp(R)$. Since R was arbitrary, U contains unstable points and we have a contradiction.

On the other hand, if $z \in J(f)$, the argument above shows that every value w of $\log z$ is unstable since the iterates of U eventually cover every compact set and hence repelling periodic points.

2.2 Classification.

The pre-periodic and periodic stable components D of a rational or entire function are completely classified: (see [**B1**])

(i) *attractive*: (resp. *super-attractive*); D lands on a component containing an attractive (super-attractive) periodic point.

(ii) *parabolic*: D lands on a component which has a parabolic periodic point on its boundary.

(iii) *rotation domain*: D lands on a component D' on which some power of the map is holomorphically conjugate to a rotation. If D' is simply connected it is a *Siegel disk*; if not it is an annulus and is a *Herman Ring* ([**H1,S3**]). Entire functions have no Herman rings but holomorphic self-maps of $\mathbb{C}^*$ may.

(iv) *essential parabolic domains*: D lands on a forward invariant domain which contains the essential singularity on its boundary and all points tend to the singularity under iteration. Rational functions have no such domains but entire functions may. We will see an example later in the paper. It is proved in [**E**] that such domains do not exist for "finite type" entire functions.

Sullivan's finiteness theorem says that every component of the stable set of a rational function is pre-periodic. His proof is based on the techniques of Teichmüller theory.

Entire functions and self-maps of $\mathbb{C}^*$ can also have wandering (non-preperiodic) domains in their stable sets. One way to construct an example of such an entire function is to begin with a function f with a forward invariant component and then to define E by $(**)$ taking an appropriate branch of the logarithm. Specific examples are shown in section 6.

The covering properties of holomorphic functions play an important role in their dynamics. ω is called a *singular value* of the function f if for some neighborhood U of ω, $f: f^{-1}(U) \to U$ is not a covering map. Isolated singular values can be critical values in the ordinary sense: that is, $\omega = f(p)$ where $f'(p) = 0$, or they can be *asymptotic values* of f, namely points for which there is a *critical path* $\alpha: [0,1) \to \mathbb{C}^*$ such that $\lim_{t \to 1} \alpha(t) = 0$ or ∞, and $\lim_{t \to 1} f \circ \alpha(t) = \omega$.

Pre-periodic components of types (i) and (ii) contain critical paths or points and singular values. Rotation domains contain the ω-limit set of a singular value in their boundary. For essential parabolic domains we have:

LEMMA 2.3. *If D is an essential parabolic domain, then it contains an asymptotic value on its boundary and it contains a critical path.*

PROOF: Assume for simplicity that D is forward invariant since we can replace the function by a power. D has an essential singularity $\alpha(= 0 \text{ or } \infty)$ as a boundary point by definition. We construct a critical path as follows. Let $z_0 \in D$ and set $z_n = f(z_{n-1})$. Join z_0 to z_1 by a path l_0. Set $l_n = f^n(l_0)$; $l_\infty = l_0 \cup l_1 \cup \cdots \cup l_n \cup \ldots l_\infty$ is forward invariant by construction and defines a critical path so ∞ is an asymptotic value.

An entire function is said to be of *finite type* if the set of singular values is finite. In [GK] we prove that every component of the stable set of an entire function of finite type is pre-periodic or periodic.

Define a holomorphic self-map of $\mathbb{C}^*$ to be of *finite type* if it has only finitely many singular values. They are isolated. It is also easy to check that finite type maps are closed under composition.

Below we will use Teichmüller theory again to prove that every component of the stable set of a finite type function is periodic or pre-periodic. It follows that the only wandering domains entire functions which are defined by such a function using $(**)$ can have, occur because of the choice of branch of the logarithm.

3. Connectivity of Stable Domains.

Let D be a component of a holomorphic self-map of $\mathbb{C}^*$. D is called *bounded* if it is bounded away from both 0 and ∞. If D is bounded away from ∞ but not from 0, call it *singly unbounded at* 0; if it is bounded away from 0 but not ∞ call it *singly unbounded at* ∞. Otherwise, call it *doubly unbounded*.

In the theory of entire maps, we distinguish between simply connected components D and multiply connected components. For the theory of holomorphic self maps of $\mathbb{C}^*$, we must make a further distinction. A doubly connected component may either separate 0 and ∞ or not. If D is doubly connected and all homotopically non-trivial curves in D separate 0 and ∞ we call D an *annule*. As we shall see, annules are in many ways similar to simply connected components.

Baker, building on the results of Töpfer [**T8,Ba2,Ba3**], developed a theory of stable components of entire functions. In this section, we develop a similar theory for maps of $\mathbb{C}^*$. The second singularity simplifies things.

Our results are summarized in:

THEOREM 3.1. *Let $f(z)$ be a holomorphic self-map of $\mathbb{C}^*$ with stable region Q, and let D be a component of Q. Then*

  (i) *D is either simply connected or an annule.*

  (ii) *If D is bounded and an annule, either it is a wandering domain and the only possible limit functions are the constants 0 and ∞, or it is a Herman ring.*

(iii) *If D is unbounded then either*

    (a) all components of Q are simply connected or,

    (b) there is a doubly unbounded annule and all other components are simply connected or,

    (c) there are exactly two singly unbounded annules, one at each singularity and all other components are simply connected or non-wandering bounded annules.

REMARK: Since our functions may have asymptotic values, it is not always true that the map is onto the component containing the image. However, for the sake of simplicity, and because it doesn't make a difference in the argument, in what follows we will write $f(C)$ both for the image of C and for the component containing the image of C under f.

We break the proof of the theorem up. Lemma 3.2 is part (i), Lemma 3.3

is part (ii) and part (iii) follows from proposition 3.4 and parts (i) and (ii).

LEMMA 3.2. *D is either simply connected or an annule.*

PROOF: Assume D is multiply connected but is not an annule. There is a compact component of the complement and a curve γ in D separating it from 0 and ∞ but not separating 0 from ∞. If B^n is the bounded component of the complement of $\gamma^n = f^n(\gamma)$, $\gamma^0 = \gamma$, $B^0 = B$, $f(B^n)$ cannot contain 0 and ∞ and so contains B^{n+1}. Since f takes compact sets in $\mathbb{C}^*$ to compact sets the iterates γ^n also do not separate 0 from ∞. Since γ is stable, a subsequence of the f^n tend to a holomorphic function on γ. However, this limit function must also be defined on all of B by a Cauchy integral. It is clearly not a finite constant since B contains Julia set. It also cannot be 0 or ∞: if it were, since the convergence is uniform, the γ^n would have to "nest" about these points, bounding neighborhoods of them, and hence separate them. It cannot be non-constant either, since if it were all the B^n would have to be contained in some compact set again contradicting the fact that B contains Julia set.

This situation is unlike that for entire functions in which distinct images of homotopically non-trivial stable curves do "nest" around infinity and the limit function is ∞. In our case "nesting" about 0 or ∞ can only occur if neither component of the complement γ is compact and D is an annule.

LEMMA 3.3. *If D is a bounded annule then it is either a wandering domain with limit functions 0 and/or ∞, or a Herman ring.*

PROOF: First assume the bounded annule D is forward invariant and let γ be a generator for it. Since $\overline{D}$ is compact in $\mathbb{C}^*$, $f: D \to D$ is proper and hence a finite ramified covering map. There cannot be ramification points however since if there were the Euler characteristic of the covering space would be negative and it is not. Since the n-fold connected cover of an annulus of modulus M is M/n, the degree of f is 1, $f: D \to D$ is a homeomorphism and so conjugate to a rotation. Since D is stable and contains no ramification points, it contains no periodic points and the rotation is irrational. If D is wandering, the iterates are again annules and a subsequence must nest about one of the essential singularities hence the limit function must be one of the constants 0 or ∞.

PROPOSITION 3.4. *There is at most one unbounded annule at each essential singularity.*

The proof follows from the following lemmas.

LEMMA 3.5. *If C is a doubly unbounded component, then no other component can be an annule.*

PROOF: Since C is doubly unbounded, given any neighborhood U of 0 and any neighborhood V of ∞, there is a curve β in C which joins a point in $U \cap C$ with a point in $V \cap C$. β can be chosen so that it intersects any generating curve of any annule. Therefore, if there is an annule, it must be C.

LEMMA 3.6. *If C is an annule which is unbounded at $\alpha(= 0, \infty)$, then there is no other annule unbounded at α.*

PROOF: Suppose C and C' are both annules unbounded at α and let γ and γ' be their respective generating curves. Since either γ or γ' intersects both C and C', $C = C'$.

PROOF OF PART (iii): If there are a wandering annule and two singly unbounded annules or a doubly unbounded annule the iterates of the generating curve of the wandering annule would eventually intersect one of the unbounded annules.

4. Teichmüller Spaces.

Let g be a quasi-conformal (qc) homeomorphism of the sphere $\hat{C}$ and f a holomorphic self-map of $\mathbb{C}^*$. Then if $h = g \circ f \circ g^{-1}$ is also holomorphic on $\mathbb{C}^*$, h is dynamically equivalent to f; that is, if z is stable for f, then $g(z)$ is stable for h and vice-versa. Even more, g conjugates the various types of stable behavior for f into the same type of stable behavior for h. We therefore define:

$$\mathcal{T}(f) = \{h \mid h = g \circ f \circ g^{-1}\}/\{\text{conjugation by } z \to az\}$$

where

$$h \text{ is holomorphic on } \mathbb{C}^*, g \text{ is } qc, g(\{0,1\}) = \{0,1\}, a \in \mathbb{C}.$$

Set $SV(f) = \{$the closure of the set of singular values$\}$. The aim of this section is to prove:

THEOREM 4.1. *If f is finite type, then $\mathcal{T}(f)$ is finite dimensional.*

We proceed as we did in [**GK**]. Let $\mu(z)$ be a measurable function on a domain $\hat{C}$ such that $\|\mu(z)\|_\infty < 1$. μ is called a *Beltrami Coefficient* of $\hat{C}$. The "measurable Riemann mapping theorem" [**AB**] says that there is a one-to-one correspondence between Beltrami coefficients μ and quasi-conformal homeomorphisms g of $\hat{C}$ which are normalized by what they do to three points. The integral, $g = g^\mu$ satisfies, $\mu(z) = g_{\bar{z}}(z)/g_z(z)$ where $g_{\bar{z}}$ and g_z are distributional derivatives. We will assume, without loss of generality that g fixes 0, 1 and ∞.

A Beltrami coefficient is *compatible* with f if $\mu(f(z))\overline{f}'(z)/f'(z) = \mu(z)$. This condition says that $h = g^{-1} \circ f \circ g$, where g is the integral of μ, is holomorphic. Define

$$T_0(f) = \{\mu \mid \mu \text{ is compatible with } f\}.$$

Define an equivalence relation on $T_0(f)$ as:

$$\nu \sim \mu \equiv_{df} g^\nu \circ (g^\mu)^{-1} \text{ is homotopic to the identity rel } SV(f).$$

Let $T(f)$ be $T_0(f)$ mod this relation. Set $M_f = \hat{C} - (\{0, \infty\} \cup SV(f))$ and $T(M_f) =$ the usual Teichmüller space of M_f; i.e.

$$T(M_f) = \{\mu \mid \mu \text{ is a Beltrami coefficient modulo the relation}$$

$$\mu \sim \nu \text{ iff } g^\mu \circ (g^\nu)^{-1} \text{ is homotopic to the identity rel}$$

$$\text{the boundary points of } M_f\}.$$

We have the following diagram:

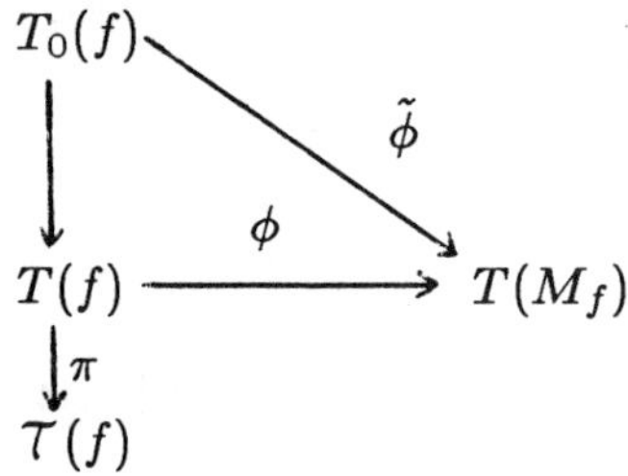

PROPOSITION 4.2. *The fiber of ϕ has complex dimension at most 2.*

PROOF: The proof is straightforward and identical to the proof of the analogous proposition in [**GK**]. It uses the fact that $f^{-1}(SV(f))$ is discrete in $f^{-1}(M_f)$ and hence that $f \colon f^{-1}(M_f) \to M_f$ is a covering map, the homotopy lifting theorem and the removable singularity theorem apply and show that if $\phi(\mu) = [\mathrm{id}] \in T(M_f)$ then g^μ is affine.

PROPOSITION 4.3. *The fiber of π is finite.*

PROOF: The proof is again straightforward and identical to the proof of the analogous result in [**GK**]. It uses the facts that the mapping class group acts discretely on $T(M_f)$, and that the group of all quasi-conformal maps commuting with f modulo the component of the identity can be naturally identified with a subgroup of the mapping class group.

Theorem 4.1 now follows since the dimension of $T(M_f)$ is finite.

5. A No Wandering Domain Theorem.

In this section we prove the following:

THEOREM 5.1. *If f is a finite type holomorphic self-map of $\mathbb{C}^*$, then f has no wandering domains.*

We will show that the existence of a wandering domain implies that $T(f)$ is an infinite dimensional space. If $SV(f)$ is finite this contradicts theorem 4.1. Our proof is based on Sullivan's argument that a rational function has no wandering domains as modified by Bers [**Be2,S1**] (see also [**GK**]).

Since $SV(f)$ is finite, if f has a wandering domain it is eventually unramified. Any stable domain has at least three boundary points and so has the unit disk U as its universal covering space. Let $D = D_0$ be an unramified wandering domain. Let $D_n = f(D_{n-1})$, $n > 0$. We have

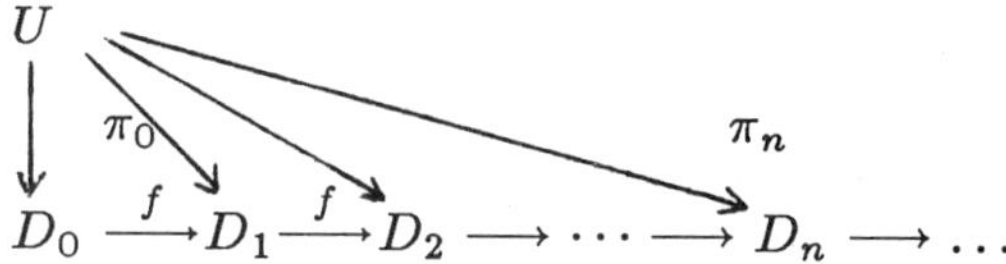

where π_i is a universal covering map and f is a covering map. Since f is unramified on D_0, the induced maps on the fundamental groups inject. We have two possibilities. By lemma 3.2 either the fundamental groups are all trivial, or there is an integer k such that the fundamental group of D_n is isomorphic to Z for all $n > k$.

The proof of theorem 5.1 follows directly from the following together with theorem 4.1.

THEOREM 5.2. *If f has a simply connected wandering domain on which it is eventually injective, then $T(f)$ is infinite dimensional.*

This theorem is proved for rational functions in [**S1**], [**Be2**] and for entire functions in [**GK**]. The proof in [**GK**] applies mutatis mutandi for holomorphic self-maps of $\mathbb{C}^*$. The delicate step is to pick a representative Beltrami differential in each equivalence class canonically.

PROPOSITION 5.3. *If $SV(f)$ is finite and D is an unramified wandering domain, then the fundamental groups of the D_n are trivial for all n.*

In order to prove proposition 5.3 we need a lemma. The basic idea in the lemma is due to Tangerman [**Ta**] who proved it for entire maps. We need first remark that if $f(z)$ has only a pole at 0, $f(z) = z^n \exp(g(z))$ where $n < 0$ and $g(z)$ is entire. It is easy to check however, that $f \circ f$ has an essential singularity at 0. The same is true if we reverse the roles of 0 and ∞. We therefore lose no generality by our assumption in the following lemma. In this lemma $SV(f)$ denotes those singular values in $\mathbb{C}^*$.

LEMMA 5.4. *Assume f has essential singularities at both 0 and ∞. Suppose $SV(f)$ is contained in a bounded annule A in $\mathbb{C}^*$. Let B be a component of $\mathbb{C}^* - A$. Then every component of $f^{-1}(B)$ is unbounded and either simply connected or an annule.*

PROOF: Let V be a component of $f^{-1}(B)$. By the remark on section 3, V is unbounded. $SV(f) \subset A$ implies $f|V : V \to B$ is a covering and the induced map on fundamental groups is injective. Since $\pi_1(B) \cong Z$, it follows that $\pi_1(V)$ is trivial or $\pi_1(V) \cong Z$. In the former case we are done. In the latter case V is homeomorphic to either a punctured disk or an annulus. If V is an annulus but not an annule, one component of the complement of the annulus is mapped onto ∞ (or 0), which is impossible for holomorphic functions. If V is a punctured disk, either the puncture is in $\mathbb{C}^*$, in which case we have a third singularity and a contradiction, or the puncture is at 0 or ∞. This means the map $f|V : V \to B$ is finite to one. This contradicts the assumption that we have essential singularities.

PROOF OF PROPOSITION 5.3: Suppose for some N, D_N is an annule for all $n \geq N$. Without loss of generality, assume $N = 0$. Let α_0 be a smooth representative for a generator of $\pi_1(D_0)$, $\alpha_n = f^n(\alpha_0)$. By lemma 3.3 D_n is bounded. Let D_n^∞ be the component of the complement of D_n containing ∞, and B_n^0 be the component of the complement of D_n containing 0. Since there are only finitely many points of $SV(f)$ in $\mathbb{C}^*$, they are all contained in a bounded annule A. There is an integer k such that for all $n > k$ either B_n^∞ (or B_n^0) contains no points of $SV(f)$. The argument in lemma 5.4

shows that each component of $f^{-1}(\alpha_n)$ is unbounded contradicting the boundedness of D_{n-1}.

6. Applications.

Knowing a parametrization of the space $T(f)$ for a given f, makes it easier to study it. The functions in the family,

$$F = \{f(z) = z^n \exp(P(z) + Q(1/z)) : n \in Z, P \text{ and } Q \text{ are}$$

$$\text{polynomial of degrees } p \text{ and } q\}.$$

are of finite type as we show in [**Ke**]. We prove there that the only singular values of these functions are their critical values and the asymptotic values 0 and ∞. In that paper we also prove:

THEOREM 6.1. *Any holomorphic function on* $\mathbb{C}^*$ *which is topologically conjugate to a member of F is holomorphically conjugate to a member.*

Corollaries of this theorem are that for any $f \in F$, the coefficients of the polynomials are parameters for $T(f)$ and that the dimension of $T(f)$ is $p + q$.

The first application we will make is to the family:

$$f_\lambda(z) = z * \exp(1/z + \lambda).$$

This is a family of functions with an essential singularity at 0 and a pole at ∞. By the above it is a one parameter family.

It is straightforward to compute the following:

(i) If Re $\lambda > 0$, all the fixed points are repelling and the (only) critical point tends to the pole at ∞. The eigenvalue there is $\exp(\lambda)$ so ∞ is an attracting fixed point. The stable domain containing this orbit is a completely invariant annule. There are no other stable components by theorems 3.1 and 5.1.

(ii) If Re $\lambda \leq 0$, the situation is more complicated; f_λ can have at most one attractive fixed point. If it does, it attracts the critical value. Since there are no wandering domains, the stable set consists only of the preimages of the component containing the fixed point. If λ is real, the negative real axis is in the Julia set and the positive real axis is stable. All stable domains are simply connected and unbounded.

20

(iii) If $\lambda = 0$, there is an essential parabolic domain whose points approach ∞. The orbit of the critical value is in the segment $(1, \infty)$ and tends to ∞. Also a small ball about 1 also approaches ∞ in a Stoltz angle under iteration so the stable set is not empty, and consists of the orbit of this domain. The negative axis is in the Julia set, so all components are simply connected.

In the above cases, the forward invariant attractive (or parabolic) domain has infinitely many prime ends. The points in the Julia set can be characterized by the symbolic dynamic techniques of Devaney and Krych [**DK**]. The Julia set contains infinitely many analytic curves described in [**DK**] as "Cantor bouquets" whose points are inaccessible from inside a stable domain (see figure 6.1).

The corresponding entire functions defined by (**) are:

$$E_\lambda(w) = w + \exp(-w) + \lambda.$$

Particular functions in this family have been studied. Fatou's original paper on entire functions [**F**] studies $E_1(w)$. He shows that there is one simply connected completely invariant domain of stability which is an essential parabolic domain. Newton's method for finding zeros of $g(w) = 1 - \exp(w)$ is the function $E_{-1}(w)$. Michel Herman used the Newton's method example to construct one of the first examples of an entire function with a wandering domain, $E_{-1+2\pi i}$.

The analysis above shows that for Re $\lambda > 0$, E_λ always has a single completely invariant simply connected essential parabolic domain. If Re $\lambda \leq 0$, all the stable domains are simply connected. They are either wandering, as in the Herman example, or they are pre-periodic, for example when λ is real. Figure 6.1 shows part of the dynamic plane for $\lambda = -1$. The figure contains the horizontal strip, $[-\pi i, \pi i]$; white is stable and black is unstable. It is symmetric about the real axis; the real line is stable. The smooth boundaries occur because of truncation. Figure 6.2 is a blow-up of part of the picture.

In figure 6.3 we plot the behavior of the dynamics of f_λ as λ varies. White regions have stable behavior, black regions don't. Since we know that the dynamics are the same for every point in the right half plane, we don't show it. Translating λ by $2\pi i n$ permutes the roles of the fixed points, so the left half of the λ-plane is periodic with period $2\pi i$. In the figure, which shows a strip of width 2π in the left half of the λ-plane, we see a Mandelbrot set.

Figure 6.1

Figure 6.2

Figure 6.3

We also see figures like those obtained by Devaney in [**DK**] for the family $\lambda \exp(z)$.

The appearance of the Mandelbrot set is not surprising. Let U be the rectangle defined by $-L \leq x \leq L'$ and $(-\pi - \varepsilon) \leq y \leq (\pi + \varepsilon)$, where L and L' are large positive constants and ε is a small positive constant. The preimage U' of U is contained inside U. (E_λ, U', U) is "polynomial-like" in the sense of Douady–Hubbard [**DH**], hence the appearance of the Mandelbrot set. The projections V and V' of U and U' by e^w are *not* simply connected on $\mathbb{C}^*$ and (f_λ, V', V) therefore does not satisfy the definition of polynomial-like. Moreover, it is impossible to find *simply connected* regions satisfying the conditions of the Douady–Hubbard definition; annules must be taken into account. In order to apply the Douady–Hubbard theory, their definition of polynomial-like must be extended to triples (f, U', U) where f is the restriction of a holomorphic self-map of $\mathbb{C}^*$ and U is either simply connected or an annule. U' must also be simply connected or an annule and have compact closure in U.

In the first family of examples, the singularity at the pole is really removable. Conjugating the functions by the map $z \to 1/z$, we obtain a family of functions, $\tilde{f}_\lambda(z) = z \exp -(z + \lambda)$, which are entire so that the new theory developed in this paper, while instructive, is not strictly necessary. The next family we study has two essential singularities so the new theory is necessary. It is the two parameter family:

$$f_{\alpha,\lambda} = z * \exp(\alpha(z + 1/z) + \lambda) \quad \alpha \in \mathbb{C}^*, \quad \lambda \in \mathbb{C}.$$

To study the dynamics for given α and λ, or to study the dependence of the dynamics on α and λ, it is easier to look at the lifted family of functions:

$$E_{\alpha,\lambda} = w + \alpha(\exp(w) + \exp(-w)) + \lambda.$$

By theorem 5.1 we know that $f_{\alpha,\lambda}$ has no wandering domains, hence any wandering domain of $E_{\alpha,\lambda}$ occurs in the lifting process.

Figure 6.4 shows the stable region (in white) and Julia set (in black) for the function $E_{1,-\sqrt{5}}(w) = w + \exp(w) + \exp(-w) - \sqrt{5}$ in the strip $(-\pi i, \pi i)$. There is a striking similarity with the Julia sets of the cosine family of functions. The large circular looking region is the forward invariant region containing the super-attractive point $w = \log((\sqrt{5} - 1)/2)$. The other regions are preimages of the super-attractive region. The picture looks

Figure 6.4: Stable region of $E_{1,-\sqrt{5}}$

essentially the same for $E_{1,-2.5}$ which has a real attractive fixed point. The real axis and the lines $y = \pm\pi i$ are invariant. There are two repelling periodic points on the real axis bounding the stable interval containing the attractive fixed point. The pre-images of these repelling fixed points are dense in the boundary of the stable region. The stable regions are therefore all bounded.

We know from theorem 6.1 that functions in $\mathcal{T}(E_{1,-2.5})$ are of the form $E_{\alpha,\lambda}$ for some subset of (λ, α) in $\mathbb{C}^2$. For each of these functions there is an invariant curve in the Julia set which is a quasiconformal image of $y = \pi i$. The components of the stable set are bounded, simply connected and deployed as they are in the figure. Again we expect to see Mandelbrot sets in the parameter space since we can find a neighborhood of the forward invariant component which is mapped over itself. The images of the components under the exponential map are components for $f_{\alpha,\lambda}$ and are not annules so the Hubbard–Douady theory applies to both families.

In the remaining figures, we plot 1-dimensional slices of the parameter space. Since there are two critical values in a strip of period 2π we test the behavior of these two points. The black regions indicate points where at least one of these critical values has a bounded orbit and the function has a non-empty stable set. Figure 6.5 is a part of the λ-plane with $\alpha = 1$ and figure 6.6 is a part of the λ-plane with $\alpha = (1 + i)/2$. Note that the Mandelbrot sets are quite evident. Figures 6.7 and 6.8 are parts of the α-plane for $\lambda = -2.5$ and $\lambda = -2.3 + .3i$ respectively. Again there are Mandelbrot sets evident here. The smooth boundary we see occurs along the imaginary axis and is probably due to the way the computer takes a branch of the arctangent.

These phenomena suggest many questions; for example, the parameter space of a cubic is also 2-dimensional and the Mandelbrot sets appear in 1-dimensional slices. How do these bifurcation sets compare with those in the cubic case? We will pursue these ideas in future work.

Department of Mathematics, Herbert H. Lehman College, C.U.N.Y., Bronx, New York 10468

Figure 6.5:
Bifurcation set of $f_{\lambda,\alpha}$ $\alpha = 1$
x-range $-4, 2$
y-range $-1.5, 1.5$

Figure 6.6

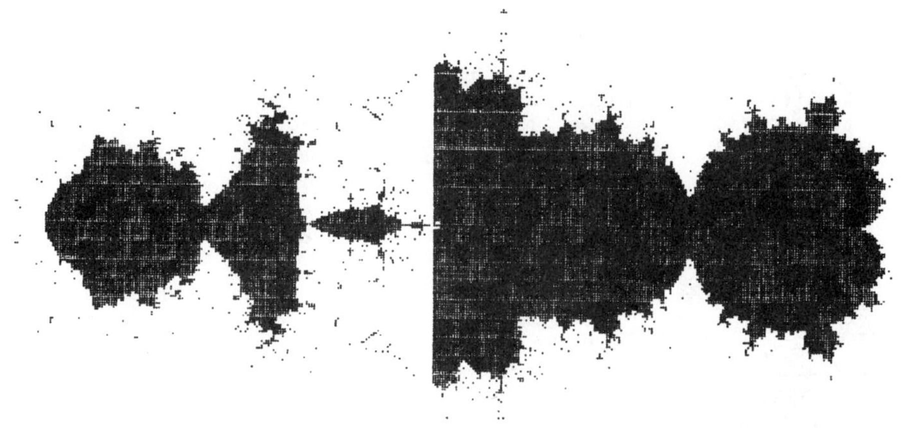

Figure 6.7:
Bifurcation set of $f_{\lambda,\alpha}$
x, y
$-2, 2 \quad \lambda = -2.5$
$-2, 2$

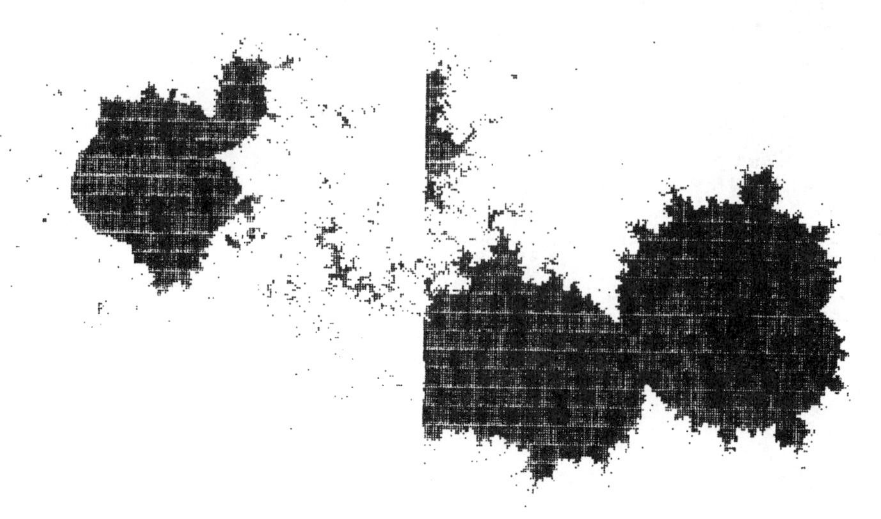

Figure 6.8:
Bifurcation set of $f_{\lambda,\alpha} \quad \lambda = -2.3 + .3i$
x, y range $\quad (-2, 2)$

REFERENCES

[A] Ahlfors, L.V., "Lectures on Quasiconformal Mappings," Van Nostrand, 1966.

[AB] Ahlfors, L.V. and Bers, L., *Riemann's mapping theorem for variable metrics*, Ann. of Math. **72** (1960), 385–404.

[Ba1] Baker, I.N., *Repulsive fixpoints of entire functions*, Math. Z. **104** (1968), 252–256.

[Ba2] ————, *The domains of normality of an entire function*, Ann. Acad. Sci. Fenn. Ser AI Math. **1** (1975), 277–283.

[Ba3] ————, *Wandering domains in the iteration of entire functions*, Proc. London Math. Soc. **49** (1984), 563–576.

[Bat] Battacharyya, P., *Iteration of analytic functions*, Ph.D. thesis, Univ. of London (1969).

[Be1] Bers, L., *Fiber spaces over Teichmüller spaces*, Acta Math. **130** (1973), 89–126.

[Be2] ————, *On Sullivan's proof of the finiteness theorem and the eventual periodicity theorem*, preprint.

[Bl] Blanchard, P., *Complex analytic dynamics on the Riemann Sphere*, Bull. A.M.S. **11** (1984), 85–141.

[DK] Devaney, R. and Krych, M., *Dynamics of* exp(z), J. Ergodic Th. and Dyn. Sys. **4** (1984), 35–52.

[DH] Douady, A. and Hubbard, J., *On the dynamics of polynomial-like mappings*, Ann. Sci. Ecole Norm. Sup. 4th Ser. **18**, 287–343.

[E] Eremenko, A., *Iteratsiitzenih funktsii*, Har'kov 1984.

[F] Fatou, P., *Sur l'iteration des fonctions transcendentes entieres*, Acta Math. **47** (1920), 337–370.

[GK] Goldberg, L. and Keen, L., *A finiteness theorem for a dynamical class of entire functions*, J. Ergodic Th. and Dyn. Sys. **6** (1986), 183–192.

[H1] Herman, M., *Sur la conjugasion différentiable des diffeomorphismes du cercle à des rotations*, Publ. I.H.E.S. **49** (1979), 5–233.

[H2] ————, *Are there critical points on the boundary of singular domains?*, Domm. Math. Phys. **99** (1985), 593–612.

[Ke] Keen, L., *Topology and growth of a special class of holomorphic self-maps of* $\mathbb{C}^*$, preprint.

[Ko] Kotus, J., *Iterated holomorophic maps on the punctured plane*, preprint.

[N] Nevanlinna, R., "Analytic Functions," Springer Verlag, 1970.

[R] Radström, H., *On the iteration of analytic functions*, Math. Scand. **1** (1953), 85–92.

[S1] Sullivan, D., *Quasiconformal homeomorphisms and dynamics I*, Acta Math. (1985).

[S2] ————, *Quasiconformal homeomorphisms and dynamics III*, preprint.

[Ta] Tangerman, F., *Personal communication.*

[Tö] Töpfer, H., *Uber die Iteration der ganzen transzendenten Funktionen insbesondere von sin z und cos z*, Math. Ann. **117** (1939), 65–84.

Automorphisms of rational maps

BY CURT MCMULLEN

Introduction.

Let $f(z)$ be a rational map, $\mathrm{Aut}(f)$ the finite group of Möbius transformations commuting with f. We study the question: when can two kinds of more flexible automorphisms of the dynamics of f be realized in $\mathrm{Aut}(g)$ for some deformation g of f?

First let $\mathrm{Mod}(f)$ denote the group of isotopy classes of quasiconformal maps commuting with f. This group acts naturally on the Teichmüller space of f. It is often infinite and quite complicated. Nevertheless, if G is any finite subgroup of $\mathrm{Mod}(f)$, it can be realized as $\mathrm{Aut}(g)$ for some deformation g of f (§ 2). (The proof uses Nielsen realization for Riemann surfaces.)

Now let $\mathrm{Mod}(J,f)$ denote quasiconformal maps only required to commute with f on its Julia set J. Elements of $\mathrm{Mod}(J,f)$ arise naturally as monodromy in bifurcation-free families of rational maps. If the Julia set is connected and carries no invariant line field, $\mathrm{Mod}(J,f)$ is a finite group (§ 3). The proof uses realization in reverse: we show $\mathrm{Mod}(J,f)$ is isomorphic to $\mathrm{Aut}(g)$ for some g.

In the analogy with Kleinian groups, $\mathrm{Aut}(f)$ plays the role of the isometry group of the quotient hyperbolic manifold, while $\mathrm{Mod}(f)$ and $\mathrm{Mod}(J,f)$ both resemble the manifold's mapping class group. Finiteness of $\mathrm{Mod}(J,f)$ might be construed as the analogue of Johansson's theorem in three dimensional topology (connectedness of J is equated with incompressible boundary (§ 4)).

Studying the realization of $\mathrm{Mod}(J,f)$ sheds light on the structure of periodic components of J. In fact if K is a component of J such that $f(K) = K$, then there is a rational map g such that the action of f on K is quasiconformally conjugate to the action of g on its Julia set $J(g)$. The proof uses 'conformal surgery' and a classification of the dynamics on the ideal boundary of K (§§ 5,6).

A perhaps unexpected subtlety is that K might not meet the boundary of any component of the domain of normality $\hat{\mathbb{C}}-J$ (see Figure I.1 and § 7).

It is conjectured that when J is connected and f is nonaffine, Mod(J,f) is always a finite group. Using a uniqueness criterion of Strebel's we establish this conjecture in the case of 'exposed critical points' (§ 8.)

In a sequel [Mc2] these finiteness results are shown to limit the possible braids arising from the motion of the attractor, which in turn leads to a topological criterion for the failure of iterative algorithms.

§ 1. Conformal Automorphisms.

Let $f(z)$ be a rational map of degree 2 or more. We denote by Aut(f) the group of Möbius transformations such that $f(Mz) = M(fz)$ for all z. For each n, M permutes the finitely many periodic points of order n; since these number more than 3 for n large enough, Aut(f) is a finite group.

If G is a subgroup of Aut(f), we can form a new rational map f/G. Namely we form the orbifold $\hat{\mathbb{C}}/G$, whose underlying manifold is again a sphere. Then f determines an endomorphism f/G of $\hat{\mathbb{C}}/G$, which is certainly conformal away from a finite set of points, and hence is again a

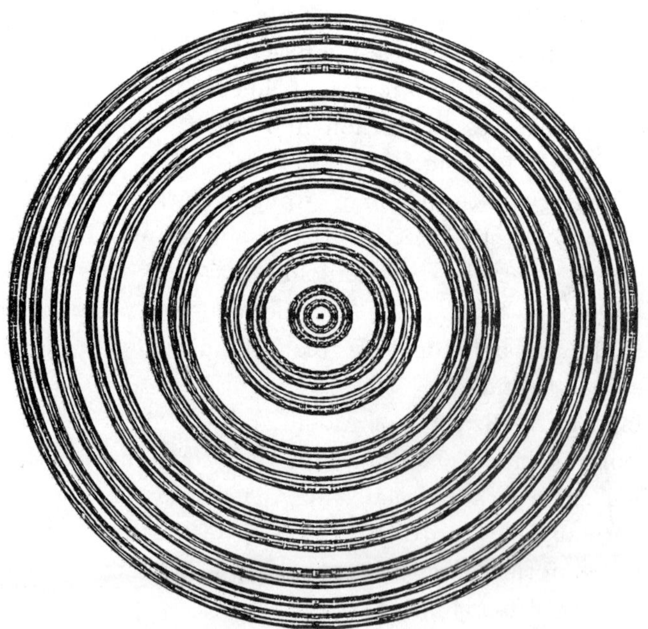

Figure I.1. The Julia set is a Cantor set of quasicircles.

rational map (of the same degree as f).

EXAMPLES:

(1) Let $f(z) = z^{n+1}$; then Aut (f) is the dihedral group generated by $z \to 1/z$ and $z \to \omega z$ where ω is a primitive nth root of unity. $T(z) = (z^n + z^{-n})/2$ gives the orbifold covering $\hat{\mathbb{C}} \to \hat{\mathbb{C}}/\mathrm{Aut}(f)$, and $T(f(z)) = g(Tz)$, where g is the quotient rational map $f/\mathrm{Aut}(f)$. If $z = e^{i\theta/n}$, then

$$T(f(z)) = \cos (n+1)\theta = g(\cos \theta),$$

so g(z) is the (n+1)st Tchebycheff polynomial.

(2) The construction works for more general Riemann surfaces. Let E be a one dimensional complex torus (elliptic curve) and let $f(z) = nz$ in the group law on E. Then Aut(f) contains the group of order two G generated by $z \to -z$, and $E/G \simeq \hat{\mathbb{C}}$ (the covering map is the Weierstrass p-function.) The map $g = f/G$ is rational and satisfies $g(p(z)) = p(nz)$.

DEFINITION: A rational map is *affine* if it is obtained as the quotient of multiplication by an integer on a complex torus as in (2).

The Julia set of an affine map is the whole Riemann sphere. These are the only known maps which carry an invariant Beltrami differential supported on the Julia set (arising from the parallel line field on E). If they are the *only* such maps, then Axiom A dynamics is open and dense in the space of all rational functions ([Sul3], [Ma]). See [L] and [Mc1] for related discussions.

Much of the sequel will be devoted to realizing less rigid automorphisms of f as elements of Aut(g) for some deformation g of f.

§ 2. Quasiconformal Automorphisms.

DEFINITION: Let QC(f) denote the group of all quasiconformal maps $\phi : \hat{\mathbb{C}} \to \hat{\mathbb{C}}$, such that $f(\phi(z)) = \phi(f(z))$ for all z. The *modular group* of f (denoted Mod(f)) is the group of isotopy classes of such ϕ: in other words, it is the quotient of QC(f) by the path component of the identity.

Mod(f) acts on Teich(f), the Teichmüller space of f, a contractible complex manifold which consists of pairs of rational functions and quasiconformal maps $<[g], [\psi]>$ such that ψ establishes a quasiconformal conjugacy between f and g $(\psi(g(z)) = f(\psi(z)))$. The brackets indicate that only the conformal conjugacy class of g and the isotopy class of ψ are remembered. The action is the obvious one: $[\phi]<[g],[\psi]> = <[g],$

$[\phi \circ \psi]>$. Any two choices for ψ differ by an element of Mod(f), so the quotient of Teich(f) by this action amounts to forgetting ψ altogether. The quotient $\mathbf{M}(f)$ is the *moduli space* of f, consisting of conformal equivalence classes of rational maps quasiconformally conjugate to f. The action of Mod(f) is properly discontinuous, so $\mathbf{M}(f)$ is a complex orbifold [Sul3].

There is an inclusion Aut(f) $\subset$ Mod(f); in fact Aut(f) is just the isotropy subgroup of $<[f],[id]>$.

THEOREM 2.1 Let G be a finite subgroup of Mod(f). Then there is a rational map g, quasiconformally conjugate to f, such that G is realized as a subgroup of Aut(g).

PROOF: We will lift the isotopy classes represented by G to a finite group of quasiconformal homeomorphisms commuting with f.

Let E denote the grand orbits of the periodic points and critical points of f. Any element of QC(f) must carry this countable dynamically-defined set into itself, and hence its values on E are constant during isotopy. Thus each isotopy class in G already determines a well-defined homeomorphism on $\overline{E}$.

By the classification of stable regions [Sul3], the complement of $\overline{E}$ can be decomposed into two totally invariant open sets, the first of which consists of punctured disks and annuli coming from Herman rings, Siegel disks and superattracting basins. Any $[\phi]$ in G commutes with a conformal S^1 action on these regions, and hence respects the rigid circular structure on their ideal boundaries. A canonical lift is determined by isotoping ϕ to be conformal on the punctured disks and an affine stretch on annuli.

The second open set collapses to a (possibly disconnected) hyperbolic Riemann surface of finite volume when divided by the grand orbit equivalence relation. G determines a finite group of isotopy classes on the quotient surface, which by Nielsen realization (Zieschang [Z, 54.3] and Kerckhoff [Ker]) can be lifted to a finite group of quasiconformal surface homeomorphisms. When pulled back to the domain of normality of f, this lift combines with the preceding choices to determines a lift of G to a finite subgroup G' of quasiconformal maps commuting with f on the whole Riemann sphere.

Since G' is finite, its elements are uniformly quasiconformal, and therefore some measurable complex structure is preserved simultaneously by G' and f (cf. [Sul1, Theorem 9]). Using the measurable Riemann mapping theorem [AB], we obtain a quasiconformal map conjugating f to a new rational function g and taking G' to a finite group of Möbius transformations commuting with g, as claimed. $\square$

DEFINITION: A rational map f(z) is *critically finite* if its critical points are preperioidic ($f^n(c) = f^m(c)$ for some n $\neq$ m.)

THEOREM 2.2 Let f(z) be a critically finite rational map. If f is not affine, then Aut(f) $\simeq$ Mod(f). In other words any quasiconformal automorphism is isotopic to a conformal automorphism.

This result is implicit in a theorem of Thurston (see [DH2]); it is worked out explicitly in [Mc1,8.1]. In particular, the modular group is finite.

REMARK: The Julia set of a critically finite map is connected. To see this, note that under iteration, every critical point outside the Julia set eventually lands on a periodic critical point. Thus the immediate basin of attraction of a periodic critical point contains no others, so it is a disk. Furthermore, each component of the complement of the Julia set eventually lands on a superattracting basin, so it admits a proper map to a disk branched over only one point (the superattracting cycle). Hence every complementary component is a disk, and the Julia set is connected.

In the following section we will see that connectedness of the Julia set already has strong consequences for automorphisms of the dynamics.

EXAMPLES:

(1) If f is affine, Mod(f) $\simeq$ PSL$_2$**Z**. (As observed by Herman [H], the linear action of SL$_2$**Z** on the torus descends to automorphisms of f).

(2) Let f(z) = $\lambda z + z^2$ for $0 < |\lambda| < 1$. The Julia set of f is a quasicircle, and the quotient Riemann surface is the union of a foliated disk and a once-punctured torus (corresponding to the basins of ∞ and 0 respectively.) Mod(f) is an infinite cyclic group generated by a Dehn twist about the unique simple closed curve on the torus which lifts to a circle enclosing 0. In fact the parameter λ establishes an isomorphism of the moduli space **M**(f) with the punctured unit disk.

PROBLEM: For f(z) a generic perturbation of $z \to z^n$, $n > 2$, the quotient Riemann surface is a pair of (n-1)-fold punctured tori (here generic means all 2n-2 critical points are simple and have distinct grand orbits). Mod(f) is a subgroup of the mapping class group of this Riemann surface; what subgroup is it?

§ 3. Stable Conjugacy.

DEFINITIONS: Let f(z), g(z) be two rational maps, not necessarily of the same degree. Let F, G be closed subsets of the Riemann sphere $\hat{\mathbf{C}}$ such that f(F) = F, g(G) = G. A map $\phi : \hat{\mathbf{C}} \to \hat{\mathbf{C}}$ determines a *stable conjugacy* (F,f) $\sim$ (G,g) if

    (1) ϕ is quasiconformal on $\hat{\mathbf{C}}$;

    (2) $\phi(F)$ = G; and

    (3) $\phi \circ f(z) = g \circ \phi(z)$ for all z in F.

By definition, two maps ϕ and ψ to determine the same stable conjugacy if they are isotopic through maps enjoying (1)-(3).

The stable self-conjugacies (F,f) $\sim$ (F,f) form a group, the *modular group of f on F*, which we denote by Mod(F,f). Of particular importance is the case F = J(f), the Julia set of f. A stable conjugacy f $\sim$ g will mean a conjugacy on their respective Julia sets. If f $\sim$ g then Mod(J(f),f) $\simeq$ Mod(J(g),g).

A family of rational maps $f_\lambda(z)$ is *holomorphic* if the coefficients of f are holomorphic functions of the parameter λ. Such families obey a strong dichotomy: either they display a rich cascade of bifurcations, or their members look rather like one another. To formulate the latter property, we say a holomorphic family is *stable* if there is a uniform upper bound on the periods of attracting cycles for maps occurring in the family. Members of a stable family need not be quasiconformally conjugate on the whole Riemann sphere, but they *are* stably conjugate on their Julia sets [MSS]. This will be made precise below.

Let X be a complex manifold parameterizing a stable holomorphic family of rational maps, including points corresponding to maps f(z) and g(z). Let A(f) denote the set of preperiodic points other than those in the Julia set.

A(f) is just the union of the grand orbits of the attracting and superattracting points and the centers of Siegel disks.

PROPOSITION 3.1 The homotopy class of a path γ connecting f to g in X determines a stable conjugacy

$$(J(f) \cup A(f),f) \sim (J(g) \cup A(g),g).$$

PROOF: By [MSS], in a stable holomorphic family of rational maps, the periodic points move without collision; the motion extends to the closure of the preperiodic points ($= J \cup A$) and commutes with the dynamics. If we follow the motion along γ from f to g we obtain a conjugacy ϕ mapping $J(f) \cup A(f)$ to $J(g) \cup A(g)$ and depending only on the homotopy class of γ.

The holomorphic motion of $J \cup A$ can be locally extended to a holomorphic motion of the whole Riemann sphere [ST], [BR], which is then quasiconformal as a function of z. The isotopy class of the extension is unique (this is true for any extension of an isotopy of a set to an ambient isotopy.)

Breaking the path γ into small segments, we see the composition of the local extensions defines a map ϕ satisfying conditions (1-4). Since the isotopy class of extension is locally well-defined, it does not change as we vary γ by homotopy; consequently the homotopy class of γ determines a stable conjugacy. $\qquad\square$

COROLLARY 3.2 Holomorphic motion around loops in X based at f determines a monodromy map

$$\pi_1(X,f) \to \mathrm{Mod}(J(f) \cup A(f),f).$$

The study of this map will be central to the discussion in [Mc2].

DEFINITION: A *line field on the Julia set* is a measurable Beltrami coefficient $\mu(z)\, d\overline{z}/dz$ supported on a positive area subset of J, such that on its support, $|\mu| = $ const. > 0. The line field is *invariant* if $f^*\mu = \mu$. The invariant Beltrami differentials supported on the Julia set form a finite dimensional vector space with a basis of invariant line fields [Sul3].

In all known examples (other than f affine), the Julia set carries no invariant line field at all.

THEOREM 3.3 Let $f(z)$ be a rational map with connected Julia set J carrying no invariant line field. Then there is a rational map g, stably conjugate to f, such that $\mathrm{Mod}(J,f)$ is realized as $\mathrm{Aut}(g)$. In particular $\mathrm{Mod}(J,f)$ is a finite group.

CONJECTURE: For any nonaffine rational map with connected Julia set,

Mod(J,f) is a finite group.

In § 8 we will establish the conjecture in the case of 'exposed critical points'.

The map g above is obtained by surgically altering f off of J so its quotient Riemann surface has no moduli. The process can even be applied to a *component* of J, yielding a more general result:

THEOREM 3.4 Let K be a component of the Julia set of f (other than a single point) such that $f(K) = K$. Then:

(a) There exists a rational map g with Julia set $J(g)$ and a stable equivalence $\phi : (K,f) \sim (J(g),g)$. The map ϕ determines an isomorphism $\mathrm{Mod}(K,f) \simeq \mathrm{Mod}(J(g),g)$.

(b) If K above carries no invariant line field, then g can be chosen so $\mathrm{Mod}(J(g),g) \simeq \mathrm{Aut}(g)$; in particular $\mathrm{Mod}(K,f)$ is a finite group.

The proof of this theorem occupies §§ 5 and 6.

COROLLARY 3.5 The Julia set of a rational map has at most countably many preperiodic components. The set of (pre)periodic points is dense in each (pre)periodic component.

PROOF: The periodic points of g are dense in $J(g)$. ◻

If we start with an expanding map, the g constructed in Theorem 3.3 will turn out to be critically finite, which together with the remark following Theorem 2.2 establishes:

COROLLARY 3.6 An expanding rational map is stably conjugate to a critically finite map if and only if its Julia set is connected.

EXAMPLES:

(1) Let $f(z) = z^n + c$, for c small. Then the Julia set of f is a quasicircle, f maps J to itself by a degree n covering, and $\mathrm{Mod}(J,f)$ is the dihedral group of symmetries of the (n-1) - gon obtained by marking the (n-1) fixed points of f in J. This group is realized as the group of automorphisms of $g(z) = z^n$, which is stably conjugate to f. Notice that g does not lie in the Teichmüller space of f; it is for this reason that we replace quasiconformal conjugacy with the weaker notion of stable conjugacy.

(2) To see the assumption of connected Julia set is needed, let $f(z) = z^3 + \lambda z^2$ for λ large. Then in [Mc1] we show the group of homeomorphisms $\mathrm{Mod}(f)|J$ (obtained by restricting each $[\phi]$ to its values on the Julia set) is infinite. Using the natural map $\mathrm{Mod}(f) \to \mathrm{Mod}(J,f)$ it follows that $\mathrm{Mod}(J,f)$ is infinite as well. (In general this map is neither an

injection nor a surjection, but it respects the values of $[\phi]$ on the Julia set.) Of course the Julia set for this map is not connected; it is a mixture of a Cantor set and a countable number of quasicircles.

(3) A simpler example is the map $f(z) = z^3 + \lambda$ for λ large. The Julia set is a Cantor set and the action of f on J is topologically conjugate to (Σ_3, σ), the one-sided shift on three symbols. $\mathrm{Mod}(J,f)|J$ is the full group of topological automorphisms of the three shift.

REMARK: When J is connected, one can construct a contractible complex manifold $\mathrm{Stab}(f)$ which parameterizes all maps stably conjugate to f. $\mathrm{Stab}(f)$ consists of pairs $<[g],[\phi]>$, where ϕ gives a stable conjugacy $f \sim g$. $\mathrm{Mod}(J,f)$ acts on this space, and the quotient furnishes a partial compactification of the moduli space of f. Thus every element of $\mathrm{Mod}(J,f)$ arises from monodromy in this universal family. One can think of Theorem 3.3 is providing a simultaneous fixed point for all elements of the modular group.

On the other hand, when J is not connected $\mathrm{Mod}(J,f)$ is always too big: there exists elements which do not arise as the monodromy in any stable family. (There are too many choices for the isotopy class of ϕ on the complement of J.) Nevertheless it is a technically convenient group to work with.

PROBLEM: Construct the stable deformation space of a rational map with possibly disconnected Julia set. Identify its modular group intrinsically in terms of f.

§ 4. Comparison with Hyperbolic Three Manifolds.

In [Sul2] a striking dictionary is established between the dynamics of rational maps and of Kleinian groups. In this section we attempt a translation of some of the preceding results. (Some *contrasts* are discussed in § 7.)

Let Γ be a finitely generated Kleinian group (discrete subgroup of $\mathrm{PSL}_2\mathbf{C}$) with limit set Λ, domain of discontinuity Ω and quotient three manifold $M^3 = (\mathbf{H}^3 \cup \Omega)/\Gamma$. For simplicity we assume M^3 is convex cocompact, i.e. geometrically finite without cusps.

The Teichmüller space of deformations of M^3 can be described as the set of pairs $<[\Gamma'], [\phi]>$, where ϕ is a quasiconformal map conjugating Γ to another Kleinian group Γ', and the brackets indicate the pair Γ', ϕ is determined only up to inner automorphism, conformal conjugacy and isotopy. Then Teich(M^3) is a contractible complex manifold, which can be identified with the Teichmüller space of the (possibly disconnected) Riemann surface Ω/Γ.

Let Mod(M^3) denote the group of quasiconformal maps conjugating Γ to itself, modulo isotopy and inner automorphism. Then Mod(M^3) is isomorphic to the mapping class group of M^3. (Here it is essential that for the mapping class group of M^3 we take autohomeomorphisms modulo *isotopy*.) The quotient of Teich(M^3) by Mod(M^3) is the moduli space of complete hyperbolic structures on M^3. (Our definitions differ slightly from those of [Kra], because we wish to identify Mod(M^3) with the mapping class group and we wish to have Teich (M^3) contractible.)

Then Theorem 2.1 above has a precise parallel (which seems to be well known):

PROPOSITION 4.1 If G is a finite subgroup of the mapping class group of M^3, then there is a quasi-isometry ϕ mapping M^3 to another complete hyperbolic manifold N^3, on which the group G is realized as isometries.

A proof can be given along the lines of Theorem 2.1, using e.g. Thurston [Th, Chapter 11] or Douady-Earle [DE] to obtain the quasi-isometry statement.

It is interesting to speculate on the connection between our finiteness result for Mod(J,f) when J is connected and known results in the topology of three manifolds. By Dehn's lemma and the loop theorem, the limit set Λ is connected if and only if M^3 has *incompressible boundary*. Thus Theorems 3.3 and 3.4 above are reminiscent of the following result of Johansson [J]:

THEOREM 4.2 If ∂M^3 is incompressible, and every essential annulus is boundary parallel, then the mapping class group of M^3 is finite.

Moreoever Thurston proved AH(M) is compact for such manifolds (see [MB]; AH(M) is the space of discrete faithful representations of Γ in the algebraic topology). One can recognize these manifolds as follows: not only is the limit set connected, but no group element has fixed points which are both on the boundary of two different components of Ω. If 'fixed points' is replaced by 'periodic point', one has a reasonable generalization to rational functions. Comparing with Thurston's theorem, we propose the following:

PROBLEM: If f(z) is a rational function with connected Julia set and no periodic point on the boundary of two components of Ω, is the closure of the moduli space $\mathbf{M}(f)$ compact in $\mathrm{Rat}_k/\mathrm{PSL}_2\mathbf{C}$?

(Here Rat_k is the space of rational maps *strictly* of degree k. The quotient space is noncompact because a map can go to infinity by dropping in degree.)

The example of a generic perturbation of z^k shows that some hypothesis beside connectedness of J is necessary.

§ 5. Conformal Surgery.

We begin the proof of 3.4 by describing the construction of isotopies and new rational functions. The gluing lemma (5.5) provides the surgical technique for producing stable conjugacies.

DEFINITION: A *disk* will mean a simply connected open subset of $\hat{\mathbf{C}}$ whose complement contains at least 2 points. In this section U and V will always denote disks. The *ideal boundary* I(U) is the space of prime ends of U. I(U) can be identified with the usual boundary $\mathbf{S}^1$ of the unit disk Δ via a Riemann mapping. The identification is well-defined up to post-composition with a Möbius automorphisms of Δ. Thus the ideal boundary has a natural projective (i.e. $\text{PSL}_2\mathbf{R}$) structure. (For example, the cross-ratio of four prime ends makes sense.)

PROPOSITION 5.1 Let $\phi : \hat{\mathbf{C}} \to \hat{\mathbf{C}}$ be a homeomorphism which is quasiconformal on a open region R and the identity on $\hat{\mathbf{C}}$–R. Then ϕ is quasiconformal on the whole Riemann sphere.

See [DH1, Lemma 2] for a more general result. The above is also stated without proof in [Kra].

PROPOSITION 5.2

(a) Let $\phi : \overline{U} \to \overline{U}$ be an orientation preserving homeomorphism which is the identity on the topological boundary of U. Then ϕ is the identity on the ideal boundary.

(b) Let $\phi : U \to U$ is quasiconformal map of U which is the identity on the ideal boundary of U. Then ϕ extends by the identity to a quasiconformal map of the whole Riemann sphere, which in particular is the identity on the topological boundary of U.

PROOF:

(a) Let C_n be a sequence of disjoint cross-cuts with distinct endpoints defining a prime end. Uniform continuity insures $\phi(C_n)$ defines another prime end, and by assumption the end points of each cross-cut are unchanged. Consider the quadrilateral Q in U bounded by $C_k \cup C_{k+1}$, for k large. Since ϕ is orientation preserving, both Q and its image lie in the same component of U–C_n. It follows that $\phi(C_n)$ defines the same prime end as the sequence C_n, so ϕ is the identity on the ideal boundary of U. (This type of argument was shown to me independently by Hinkkanen and Douady.)

(b) A quasiconformal map which is the identity on the ideal boundary moves points only a bounded distance in the Poincare' metric on U. Since the Poincare' metric tends to ∞ relative to the spherical metric on U, ϕ extends continuously to the identity on ∂U. Suppose we extend ϕ by the identity over the remainder of the sphere. The result is a homeomorphisms which is quasiconformal on U and the identity on $\hat{\mathbf{C}}$–U; so we may apply Proposition 5.1. $\qquad\square$

PROPOSITION 5.3   Let ϕ, ψ : $\overline{U} \to \overline{V}$ be continuous maps, K-quasiconformal on U and equal on ∂U. Then there exists an isotopy ϕ_t connecting ϕ to ψ through K^3-quasiconformal maps, all agreeing on ∂U.

PROOF: Let $\alpha = \psi^{-1} \circ \phi : U \to U$. Then (i) α is K^2-quasiconformal and (ii) α equals the identity on the ideal boundary of U. We can construct an isotopy α_t connecting α to the identity map through homeomorphisms enjoying these two properties. Just uniformize U and bring the identity in radially from the boundary of the unit disk.

By 5.1 each α_t extends to the identity on ∂U. The desired isotopy is given by $\phi_t = \psi \circ \alpha_t$. $\qquad\square$

COROLLARY 5.4   If K is connected and periodic points of f are dense in K, then any $[\phi]$ in Mod(K,f) is uniquely determined by its values on K.

REMARK: Density of periodic points implies the value of $[\phi]$ on K is well-defined, since any isotopic map must agree with ϕ on periodic points.

PROOF: Suppose $[\phi]$ is the identity on K. Then ϕ is the identity on the ideal boundary of every disk in the complement of K. By the above propositions, on each disk there is an isotopy connecting ϕ to the identity through quasiconformal maps which are the identity on K. The dilatation of the intervening maps is bounded by $K(\phi)^3$ independent of the disk. By 5.1 the time t maps in each disk piece together to form a quasiconformal map of the whole Riemann sphere. Thus $[\phi] = [\text{id}]$ in Mod(K,f). $\qquad\square$

We now turn to the Gluing Lemma, which allows one to redefine a rational map f(z) on a disk U using a new map which is compatible on with f on the ideal boundary.

DEFINITIONS: Let $K \subset \hat{\mathbf{C}}$ be an infinite continuum (i.e. a continuum which is neither empty nor a single point). Then every component of $\hat{\mathbf{C}}$–K is a disk and we define I(K), the *ideal boundary of K*, to be the union of the ideal boundaries of its complementary components.

Let f be a rational map such that $f^{-1}(K)$ consists of K itself and finitely many other disjoint components. (In applications, K will be a component of the Julia set such that $f(K) = K$).

Then the map f gives rise to a real-analytic map f: I(K) → I(K) as follows. Let U be a component of $\hat{\mathbf{C}}$–K, and let U' denote the component of U–f^{-1}K whose closure contains ∂U. Then f carries U' to V by a proper map, where V is another component of $\hat{\mathbf{C}}$–K. On the uniformizations of U and V, f carries a neighborhood of $\mathbf{S}^1$ to a neighborhood of $\mathbf{S}^1$. By Schwarz reflection, f extends to a real-analytic map from I(U) to I(V).

Note that we *cannot* assume that f(U) = V and in fact f(U) may cover the whole Riemann sphere many times.

The immediate result of surgery will not be a rational map, but a more flexible object:

DEFINITION: A *quasirational function* is a branched covering f : $\hat{\mathbf{C}} \to \hat{\mathbf{C}}$ which is

  (1) locally conjugate to a holomorphic map after quasiconformal change of coordinates in the domain and range; and

  (2) whose iterates f^n are K-quasiconformal (away from branch points) for some K independent of n.

By [Sul1, Theorem 9], a function is quasirational if and only if it is quasiconformally conjugate to a rational function.

To allow us to carry out a succession of surgeries without constantly converting the results back to rational functions, we will state the gluing lemma for quasirational functions. Using Sullivan's theorem, it is easy to check that the map f: I(K) → I(K) makes sense when f is only quasirational; however the map on ideal boundary is then only quasisymmetrically conjugate to a real-analytic map.

PROPOSITION 5.5 (The Gluing Lemma): Let f(z) be a quasirational function, and let K be an infinite continuum such that $f^{-1}(K)$ consists of K itself and finitely many other disjoint components. Let B : $\Delta \to \Delta$ be a proper holomorphic map (a finite Blaschke product).

(a) Suppose f: $I(U) \to I(U)$ for some component U of $\hat{\mathbb{C}}-K$, and $\phi : U \to \Delta$ is a quasiconformal mapping whose ideal boundary values $I(U) \to \mathbf{S}^1$ give a conjugacy between the action of f on the ideal boundary of U and that of B on $\mathbf{S}^1$. Then the function

$$g(z) \;=\; f(z) \qquad\qquad \text{for z not in U}$$

$$\;=\; \phi^{-1} \circ B \circ \phi(z) \quad \text{for z in U}$$

is quasirational.

(b) Suppose U and V are components of $\hat{\mathbb{C}}-K$ such that f: $I(U) \to I(V)$ and $f^n I(U) \cap I(U) = \emptyset$ for all $n > 0$. If ϕ, ψ are quasiconformal maps of U and V to Δ such that $\psi \circ f = B \circ \phi$ on the ideal boundary of U, then

$$g(z) \;=\; f(z) \qquad\qquad \text{for z not in U}$$

$$\;=\; \psi^{-1} \circ B \circ \phi(z) \quad \text{for z in U}$$

is quasirational.

NOTATION: For part (a) we will write $g = f \cup_{\phi} B$, for part (b) $g = f \cup_{(\phi,\psi)} B$.

PROOF: In either case, consider $f^{-1} \circ g(z)$ locally near ∂U. This map is the identity outside of U, quasiconformal on U and the identity on the ideal boundary of U. Therefore it is locally quasiconformal. Composing with f, we conclude g is locally quasiconformal away from branch points. Moreover the dilation of g is bounded by the supremum of $K(f)$ and $K(\phi)^2$.

In case (a), $g^n = f^n \cup_{\phi} B^n$, so $K(g^n)$ is bounded by max $(K(f^n), K(\phi)^2)$. Since f is quasirational, the dilatation of its iterates is uniformly bounded, so the same is true for g. In case (b) we use the non-recurrent assumption on U to check the uniform bound on $K(g^n)$. $\square$

REMARKS:
(1) The degree of f may be decreased by surgery.
(2) This type of construction was invented by Douady and Hubbard to realize a polynomial-like map as a polynomial [DH1].

§ 6. Dynamics on the Ideal Boundary.

We now turn to the action of f on the ideal boundary of a component K of its Julia set, and apply the gluing procedures introduced above to complete the proof of realization for Mod(K,f) (Theorem 3.4; see 6.10 below).

DEFINITION: We define three *rigid models* for proper self-maps of the unit disk Δ:

(a) the *elliptic model* $z \to e^{2\pi i\theta}z$ where θ is irrational;

(b) the *hyperbolic model* $z \to z^d$ for some $d > 1$; and

(c) the *parabolic model* $z \to P_d(z)$ for $d > 1$, where

$$P_d(z) = \frac{z^d + a}{1 + az^d}, \quad \text{for } a = (d-1)/(d+1)$$

is the unique map of degree d with a single critical point at $z=0$ and an indifferent fixed point of multiplicity three at $z=1$.

REMARKS:

(1) Let B(z) be one of the above models, $\phi : S^1 \to S^1$ a quasisymmetric automorphism of B ($\phi B = B\phi$). Then ϕ is actually a rotation $z \to \alpha z$, $|\alpha| = 1$, and $B(\alpha z) = \alpha B(z)$ throughout the disk. It is for this reason we call these models rigid. For the elliptic model α can be arbitrary; for the hyperbolic model we must have $\alpha^d = 1$; and in the parabolic case $\alpha = 1$ (the only automorphism is the identity.)

(2) One may demonstrate these assertions for models (a) and (b) by lifting to the universal cover of S^1; in fact the only *topological* (orientation preserving) automorphisms of (a) and (b) are the rotations just described. Thus a topological map commuting with z^d is determined by its value on any one of the (d-1) fixed points.

(3) The parabolic model (c) is topologically conjugate to the hyperbolic model (b) of the same degree, so its topological automorphism are also determined by their value on a fixed point.

(4) There is no quasisymmetric conjugacy carrying the indifferent fixed point of P_d to a repelling fixed point of any other map, since a repelling point expands small neighborhoods at a geometric rate and an indifferent point does not. Thus a quasisymmetric automorphism of P_d must fix its indifferent fixed point on S^1 (the remaining fixed points are repelling) and, since an automorphism is determined by the image of a fixed point, it must be the identity.

(5) Similarly, a topological conjugacy between z^d and P_d cannot be quasisymmetric. Thus different rigid models are not quasisymmetrically conjugate on $\mathbf{S}^1$.

Now let $f(z)$ be a rational map, and let K be an infinite component of its Julia set J such that $f(K) = K$. Then $f^{-1}(K)$ consists of a finite union of components of J, one of which is K itself. Thus f gives rise to map $f: I(K) \to I(K)$ on the ideal boundary of K, as discussed in §5. Our first goal is to study the dynamics of f on the ideal boundary of K.

THEOREM 6.1

(a) Every component of $I(K)$ is preperiodic, and there are only finitely many periodic components.

(b) On each periodic component of the ideal boundary, the first return map is quasisymmetrically conjugate to one of the rigid models.

(c) All but finitely many components of $\hat{\mathbf{C}}$–K map to their images by conformal isomorphism.

(d) If C is a component of $I(K)$ which is not periodic, then there are quasisymmetric maps ϕ and ψ carrying C and $f(C)$ to $\mathbf{S}^1$ and conjugating f to $z \to z^d$ for some d (possibly equal to 1).

The proof of (a,c and d) is fairly short; the proof of (b) occupies Propositions 6.2-6.8, which give a case-by-case study of the way in which the dynamics of f on K fits in with its global dynamics.

PROOF 6.1(c) and (d): Let U be a component of $\hat{\mathbf{C}}$–K. Notice that $f^{-1}(K)$ consists of finitely many components of J (one of them K itself). If U meets neither $f^{-1}(K)$ nor contains any of the finitely many critical points of f, then as claimed in (c), f carries U to another component of $\hat{\mathbf{C}}$–K by conformal isomorphism (since it is a proper local homeomorphism of a disk).

On the other hand, if U is one of the finitely many components which contain a critical point or meet $f^{-1}(K)$, we can still find an annulus U' $\subset$ U such that $\partial U \subset \partial U'$ and f maps U' to $f(U')$ by a covering. In the coordinates on ideal boundaries obtained by *separately* uniformizing U' and $f(U')$ as standard annuli of the form

$$\{z : r < |z| < 1\},$$

f is conjugate to $z \to z^d$ for some d; but these coordinates differ only real-analytically from the standard coordinates on $I(K)$, establishing (d).

$\square$

PROOF 6.1(a): Eventual periodicity and finiteness of the number of periodic cycles for the action of f on I(K) is an immediate consequence of the following:

PROPOSITION 6.2 Let U be a component of $\hat{\mathbb{C}}$–K. Then I(U) eventually maps to I(V) where V is either a Siegel disk for f or one of the finitely many components of $\hat{\mathbb{C}}$–K containing a critical point or meeting $f^{-1}(K)$.

PROOF: If $f^n(I(U))$ never encounters a critical point or $f^{-1}(K)$, then as discussed above f^n maps U by conformal isomorphism for all n. Then $f^n(U)$ must eventually cycle, for otherwise f would have a wandering domain (ruled out by [Sul2]). Clearly $<f^n>$ is a normal family on U, so U eventually maps onto a stable region for f on which the first return map is a conformal isomorphism of a disk. Such a stable region must be a Siegel disk. □

It remains only show the periodic components of I(K) behave up to quasisymmetric conjugacy like one of the rigid models.

DEFINITION: A real-analytic endomorphism h : $\mathbf{S}^1 \to \mathbf{S}^1$ is *expanding* if there is a smooth metric $\rho(\theta)d\theta$ with respect to which the derivate of h is everywhere > 1. The model example is $z \to z^d$, d > 1.

PROPOSITION 6.3 Any expanding $C^{1+\alpha}$ orientation-preserving endomorphism of $\mathbf{S}^1$ is quasisymmetrically conjugate to $z \to z^d$ for some d > 1.

PROOF: This is a special case of two general facts: expanding maps on compact manifolds are classified up to topological conjugacy by their action on the fundamental group [Shub], and topological conjugacies between expanding conformal dynamical systems are quasiconformal [Sul3]. □

For the remainder of this section, U will be a component of $\hat{\mathbb{C}}$–K such that I(U) is periodic. Replacing f by an iterate we will assume I(U) is fixed. As discussed above, there is an annulus U' $\subset$ U on which f is a covering map to its image.

PROPOSITION 6.4 If U' can be chosen so it is a proper subset of f(U') then f is quasisymmetrically conjugate on I(U) to $z \to z^d$ for some d > 1.

PROOF: Uniformizing U, f becomes a map defined near the boundary of the unit disk which carries an annular neighborhood of $\mathbf{S}^1$ properly over itself. By Schwarz reflection we obtain a holomorphic covering F : A $\to$ B, where A and B are two annuli containing $\mathbf{S}^1$ and A is a proper subset of B. It follows that F is expanding on $\mathbf{S}^1$ relative to the

Poincare' metric $\rho(z)\,|dz|$ on B. (This argument is identical to the discussion of the *external map* in [DH1].) $\qquad\square$

Let Ω denote $\mathbf{\hat{C}}$–J. The next Proposition reduces the proof of Theorem 6.1 to the case that ∂U is part of the boundary of a periodic component of Ω.

PROPOSITION 6.5 Either ∂U is contained in ∂V for some component V of $\Omega\cap U$, or $f : I(U) \to I(U)$ is quasisymmetrically conjugate to $z \to z^d$.

REMARK: We will see in § 7 that it is actually possible for K to be enclosed in a nest of components of J, so that K (and hence ∂U) does not meet the boundary of any component of Ω.

PROOF: Since K is a component of J, there exist disjoint simple closed curves γ_n lying in $U\cap\Omega$ such that γ_n lies in an r_n neighborhood of ∂U, where $r_n \to 0$ (apply Zoretti's theorem [W, p.109]). If ∂U is not contained in ∂V for some component V of Ω, then these curves lie in infinitely many distinct components of Ω.

Pick γ_n sufficiently large that the annulus V' bounded by ∂U and γ_n contains no critical values of f, and such that γ_n does not lie in a periodic component of Ω (this is possible because there are only finitely many periodic components.) Let U' denote the (unique) component of $f^{-1}(V')$ which lies in U, such that U' is bounded by ∂U and some pre-image of γ_n. Since γ_n is disjoint from its pre-image (by virtue of lying in an aperiodic component), either $U' \subset V'$ or $V' \subset U'$ (equality cannot hold). But the modulus of U' is less than or equal to that of V', so U' is a proper subset of $V' = f(U')$. The preceding proposition applies to complete the proof. $\qquad\square$

We may now assume that $\partial U \subset \partial V$ for some component V of $\Omega\cap U$. (We may *not* assume U=V and indeed there may be regions in U whose images cover the whole Riemann sphere.) Using the classification of stable regions [Sul3], the following three propositions complete the proof of Theorem 6.1.

PROPOSITION 6.6 If V is a Herman ring or a Siegel disk, then $f : I(U) \to I(U)$ is real-analytically (and hence quasisymmetrically) conjugate to an irrational rotation $z \to e^{i\theta}z$.

PROOF: If V is a Siegel disk then $U = V$ and f is already conformally conjugate to an irrational rotation of the uniformization of V. If V is a Herman ring then f is conformally conjugate to an irrational rotation of the standard annulus uniformizing V, one of whose ideal boundary components differs from that of U only by a real-analytic transformation. $\square$

PROPOSITION 6.7 If V is an attracting or superattracting basin, or a parabolic basin whose indifferent fixed point is not contained in ∂U, then f : $I(U) \to I(U)$ is quasisymmetrically conjugate to $z \to z^d$ for some $d > 1$.

PROOF: Let $W \subseteq V$ be a Jordan domain such that f(W) is a proper subset of W and $\overline{W}$ does not meet ∂U. In the attracting and superattracting cases, we may take W to be a small round disk centered at the attracting fixed point in V. In the parabolic case, take W to be a horoball-like region (a small petal or sepal of the 'Fatou flower') such that $\overline{W}$ meets ∂V only at the indifferent fixed point.

Let W_n denote the component of the nth pre-image of W which contains W. Then

$$\overline{W}_1 \subset \overline{W}_2 \subset ... \subset U \quad \text{and:} \quad \cup W_n = V.$$

(In the parabolic case, since $f(\partial U) = \partial U$, the boundary of U avoids the grand orbit of the indifferent fixed point.)

For each n, there is unique component of ∂W_n which together with ∂U bounds an annular region A_n disjoint from W_n (Figure 6.1).

A_n lies within an r_n neighborhood of ∂U, where $r_n \to 0$; otherwise there is a continuum in ∂V which lies in U and meets ∂U, contradicting the assumption that K is a component of the Julia set.

Choosing n large enough, we can insure that A_n contains no critical

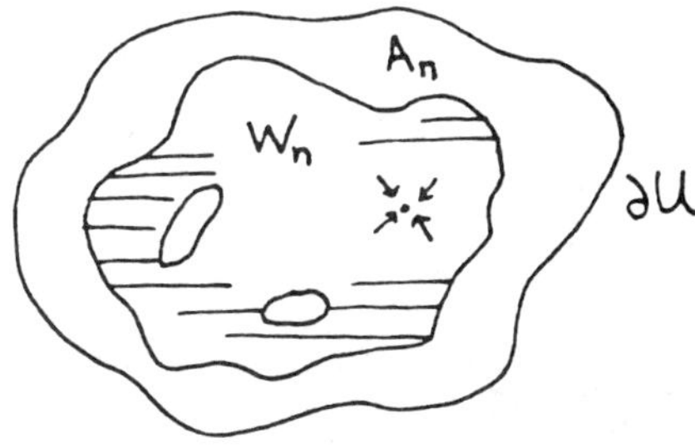

Figure 6.1. Locating an expanding annulus.

values of f. Then A_{n+1} is a component of $f^{-1}(A_n)$, as well as a proper subset of A_n, so Proposition 6.4 applies to complete the proof. $\qquad\square$

PROPOSITION 6.8 If V is a parabolic basin, with indifferent fixed point contained in ∂U, then f: $I(U) \to I(U)$ is quasisymmetrically conjugate to $z \to P_d(z)$ for some d > 1, where P_d is the parabolic rigid model.

PROOF: We begin by uniformizing U. Then f and V carry over to the unit disk Δ, where we continue to denote them by f and V. The map f is is defined only on a subset of the disk, but this subset includes V as well as an annular neighborhood of $\mathbf{S}^1$. By Schwarz reflection f is analytic on a neighborhood of $\mathbf{S}^1$, and has a unique indifferent fixed point on $\mathbf{S}^1$, corresponding to the indifferent fixed point in ∂U. We may normalize so this fixed point is at $z = 1$.

Let $W \subset V$ be a small horoball touching the circle at the fixed point. For simplicity choose W such that its boundary avoids the grand orbits of the critical points. As before, we let W_n denote the component of $f^{-n}(W)$ which contains W. There is a unique component F_n of ∂W_n lying between ∂U and W_n, and meeting ∂U at the nth preimages of 1. For n sufficiently large, F_n lies in the annulus near the boundary of Δ on which f is a covering.

Carrying out the same discussion with f replaced by P_d (where d = the degree of f on $\mathbf{S}^1$), we obtain a similar picture in another copy of the unit disk. Construct a quasiconformal map ϕ carrying the 'fundamental domain' A bounded by F_n and F_{n+1} to the corresponding region for $P_d(z)$

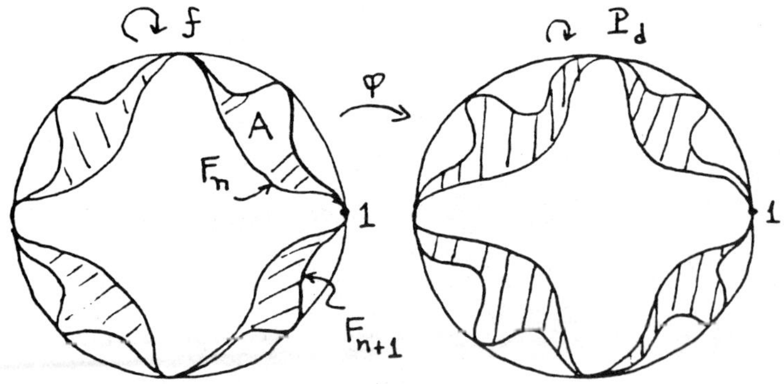

Figure 6.2. The parabolic case.

(Figure 6.2). We can arrange that the map is Lipschitz and commutes with f on the boundary of A.

Now use the dynamics on both sides to extend ϕ to a conjugacy between f and P_d on the pinched annulus bounded by F_n and S^1. The resulting map commutes with f on this pinched region and has a quasiconformal extension to the full disk because of our Lipschitz requirement. Thus it induces a quasisymmetric conjugacy on S^1, the ideal boundary of U. $\qquad\square$

PROOF 6.1(b): Let I(U) be a periodic component of the ideal boundary of K. Apply Proposition 6.5 if ∂U is not contained in ∂V for V a component of $\Omega \cap U$; otherwise V is a Siegel disk, Herman ring, or an attracting, superattracting or parabolic basin; apply 6.6, 6.7 or 6.8 as appropriate. $\qquad\square$

Using the preceding analysis of the dynamics on the ideal boundary of K, we may now apply surgery to simplify the dynamics of f in the complement of K by pasting in rigid models.

PROPOSITION 6.9 There is a stable conjugacy $(K,f) \sim (J(g),g)$, where $g(z)$ is a rational map with the following properties. The Julia set $J(g)$ is connected, and the stable regions of g are Siegel disks, superattracting and parabolic basins. On each stable region the first return is conformally conjugate to one of the three rigid models. If c is a critical point in $\hat{\mathbb{C}}-J(g)$ which does not lie in a stable region, then there is a least n such that $g^n(c) \in U$ and U is a stable region; and

(a) if U is a Siegel disk, $g^n(c)$ coincides with the indifferent period cycle which forms the center of the disk;

(b) if U is a superattracting basin, $g^n(c)$ lands on the superattracting periodic cycle; and

(c) if U is a parabolic basin, $g^{n+k}(c)$ is also a critical point, where $k \geq 0$ is the least integer such that $g^k(U)$ contains a critical point.

REMARKS:

(1) It follows that the quotient Riemann surface S(g) consists of foliated disks and triply-punctured spheres. In particular it has no moduli.

(2) When there are parabolic regions, one must take care in constructing g(z) since the critical point in the periodic basin cannot be made periodic. This is the reason for the convention in (c). Without such care it may not be true that every element of Mod(J(g),g) is isotopic

to a Möbius transformation.

(3) The degree of g is > 1. This is implicit in our assumption that K (and hence J(g)) is not a single point.

PROOF: We construct a quasirational map q(z) by applying the gluing lemma finitely many times to components of the complement of K.

Consider first components U such that I(U) is periodic (by 6.1(a), there are only a finite number of such U). For simplicity assume that I(U) is actually fixed by f. Using the part (a) of the Gluing Lemma, define g(z) on U by pasting in the rigid model to which f is quasisymmetrical conjugate by Theorem 6.1(b).

Now work backwards along the pre-images of the periodic components of I(K), pasting in copies of $z \to z^d$ using part (b) of the gluing lemma and 6.1(d). Whenever $d > 1$, choose the maps ψ and ϕ such that under iteration the critical point lands on 0 in the standard coordinate systems for the rigid models. In other words, assure that the preperiodic critical points introduced land on the center of a Siegel disk in case (a), the superattracting periodic point in case (b), and the unique critical point in the immediate basin in case (c).

Once we have dealt with f^{-k} of the periodic components of I(K) for k large enough, f maps all the remaining components of $\hat{\mathbb{C}}-K$ by conformal isomorphism to their images. At this point we stop altering f.

The result is a quasirational map q(z), agreeing with f on K, such that $q^{-1}(K) = K$. The map q(z) is quasiconformally conjugate to a rational map g(z), so we have a stable conjugacy $(K,f) \sim (F,g)$, where F is a connected nowhere dense totally invariant set for g.

We must show $F = J(g)$. It is enough to show the boundaries of the periodic components of $\hat{\mathbb{C}}-F$ are in J(g), since their grand orbits are dense in F. This is clear for the hyperbolic and parabolic models; for the elliptic case, we need only rule out the possibility that g(z) is an irrational rotation of the whole Riemann sphere, and F is an invariant circle. This could only arise if K is a quasicircle bounding a Siegel disk or Herman ring on both sides (as we have seen in 6.5-6.8). But then the iterates of f form a normal family on an annular neighborhood of K, so K is not a component of the Julia set. $\qquad\qquad\square$

Since all automorphisms of the rigid models are essentialy conformal, we have good control over the stable automorphisms of (J(g),g). This is made precise by the following proposition, which completes the proof of Theorem 3.4.

PROPOSITION 6.10 The stable conjugacy $(K,f) \sim (J(g),g)$ establishes an isomorphism $\mathrm{Mod}(K,f) \simeq \mathrm{Mod}(J(g),g)$. Every $[\phi]$ in the latter group has a unique representative ϕ which is conformal off the Julia set and commutes with g on the whole Riemann sphere. If K carries no invariant line field, ϕ is globally conformal and $\mathrm{Mod}(J(g),g) \simeq \mathrm{Aut}(g)$.

REMARK: Even without line field assumptions, the representative ϕ establishes an isomorphism $\mathrm{Mod}(J(g),g) \simeq \mathrm{Mod}(g)$.

PROOF: That the conjugacy gives an isomorphism is immediate from the definitions.

Let $[\phi]$ be an element of $\mathrm{Mod}(J(g),g)$. We claim the ideal boundary values of ϕ on any component of $I(J(g))$ coincide with those of a conformal map commuting with g. If so, then by Cor. 5.4 there is an isotopy connecting ϕ to a unique quasiconformal map which is conformal on $\Omega = \hat{\mathbb{C}} - J(g)$ and commutes with g everywhere. But if $J(g)$ carries no line field, such a map is conformal, and hence $[\phi]$ is represented by an element of $\mathrm{Aut}(g)$, completing the proof.

To establish the claim, we begin by replacing g with a suitable power so that all the periodic components of Ω are actually fixed. By construction, the fixed components of Ω are either Siegel disks, parabolic basins or superattracting basins, which are conjugate to rigid models after uniformization. On the ideal boundary of each fixed component, ϕ establishes a quasisymmetric conjugacy between two of the models.

As remarked earlier, the two models must be identical, and up to composition with uniformizing maps, ϕ is a rotation commuting with the dynamics throughout the disk. Thus the ideal boundary values of ϕ coincide with those of a conformal map commuting with g.

Now let U be a preperiodic component of Ω, and let V be $\phi(U)$. Then there is a least integer k such that $g^k(U)$ (and consequently $g^k(V)$) are components fixed by g. If g^k maps U by degree 1, then the same is true for V and $g^{-k} \circ \phi \circ g^k$ extends to a conformal isomorphism from U to V with the same ideal boundary values of ϕ as claimed.

On the other hand, if U and V map by degree > 1, then by construction g^k (when restricted to U or V) has a unique critical value, which either coincides with the center of a Siegel disk or the critical point in a parabolic or superattracting basin. Then the conformal extension of the boundary values of ϕ to a map from $g^k(U)$ to $g^k(V)$ carries the critical value to the critical value, and hence there is a lift $g^{-k} \circ \phi \circ g^k$ which is a conformal isomorphism between U to V and whose boundary values agree with those of ϕ.

Thus on each component of Ω, ϕ can be replaced by a conformal map commuting with g. $\qquad\square$

§ 7. Example: A Julia Set with a Buried Quasicircle.

We will describe a Julia set with a component K such that $f(K) = K$ but K does not meet the boundary of any component of Ω. We refer to such a K as a *buried* component. For comparison we recall some topological properties of Kleinian groups.

PROPOSITION 7.1 Let Γ be a finitely generated Kleinian group with limit set Λ and domain of discontinuity Ω. Then:

(a) The diameters of the components of Ω tend to zero.

(b) Any buried component of Λ is a single point.

(c) Fixed points of elements of Γ are dense in any component of Λ which is not a point. (In particular any such component is the limit set of its stablilizer.)

(d) Under the action of Γ, the components of Λ which are not points fall into finitely many equivalence classes.

Part (a) is proved by Maskit [Mskt]; parts (b-d) can be verified inductively using the decomposition theorem [AM]. In [A] Abikoff gives an example of a limit set with a buried point (as in (b) above).

In our example the analogues of (a-d) fail to hold. As usual, J will denote the Julia set, Ω the domain of normality $\hat{\mathbb{C}}$-J.

PROPOSITION 7.2 Let $f(z) = z^2 + \lambda/z^3$. For all λ sufficiently small (but $\neq 0$),

(1) The diameters of the components of Ω do not tend to zero.

(2) The Julia set J is homeomorphic to [a Cantor set] x S^1.

(3) Every component of J is a quasicircle.

(4) J has uncountably many wandering components.

(5) J has uncountably many buried components.

(6) There is a unique buried component such that $f(K) = K$.

REMARK: (6) shows the case dealt with by Proposition 6.5 can actually occur.

The Julia set for this map, with $\lambda = 10^{-9}$, is rendered in the illustration at the beginning of this paper.

The above properties are easily verified from a picture of the dynamics of f. When λ is small, there is a large annulus A centered at zero,

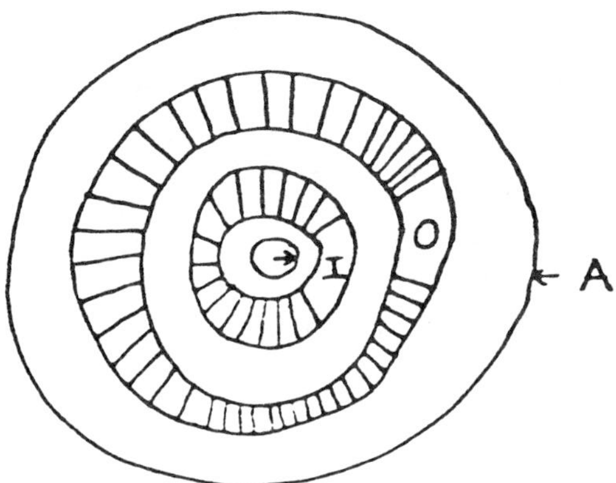

Figure 7.1. An expanding map on nested annuli.

containing no critical values, and whose pre-image consists of an inner annulus I and an outer annulus O, both nested inside of A. O maps to A by degree 2 (by a small perturbation of $z \to z^2$), and I maps to A by degree 3 (by a small perturbation of $z \to \lambda/z^3$) (Figure 7.1). The inner region enclosed by I maps to a neighborhood of infinity. Points not in A (such as the critical values of f) tend to infinity under iteration. In particular f is expanding.

Note that to construct this picture conformally, it is necessary that the annuli I and O fit inside of the large annulus A; we need $\mathrm{mod}(I) + \mathrm{mod}(O) \leq \mathrm{mod}(A)$ (and an easy argument rules out equality if the picture occurs inside of a rational map). Since I and O are coverings of A by degrees 2 and 3, and $1/2 + 1/3 < 1$, the picture is possible. The map $z^n + \lambda/z^m$ gives a similar example whenever $1/n + 1/m < 1$; however we do not know a simple example of a buried component for a rational map of degree less than 5.

The Julia set can be constructed as $J = \cap f^{-n}(A)$. Using this construction, one arrives at a description of the topological dynamics. The components of the Julia set are circles nested between 0 from ∞. This gives an ordering to the components: $K < L$ if K separates L from 0. The set of components is isomorphic to the one-sided shift on two symbols

$$\Sigma_2 = \{0,1\}^{\mathbf{N}}$$

with the lexicographical ordering. With this identification, f determines a 2-to-1 endomorphism σ of Σ_2, namely

$$\sigma(a_1, a_2, \cdots) = (a_2, a_3, \ldots) \qquad \text{if } a_1 = 1,$$

$$= (1-a_2, 1-a_3, \cdots) \quad \text{if } a_1 = 0.$$

Thus f has exactly two fixed components, coded $(1,1,1,1\ldots)$ and $(0,1,0,1,0,\ldots)$. The former is the outermost circle and the latter is the unique buried component K such that $f(K) = K$.

The Julia set as a whole is homeomorphic to $\Sigma_2 \times \mathbf{S}^1$. Identifying $\mathbf{S}^1$ with the unit circle in the complex plane, the topological dynamics is the skewed product

$$F(<a_i>, z) = (\sigma <a_i>, z^{5a_1-3}).$$

In other words F maps the circles in the outer annulus by degree 2 and those in the inner annulus by degree -3. The expanding property implies the circles in J are actually quasicircles.

It is not hard to see how the surgery procedure of § 6 applies to the dynamics of f on the component K. Since K is buried, one can find a small annulus enclosing K whose boundary consists of two other components of the Julia set. This annulus is mapped properly over itself, so f is expanding on the ideal boundary of K. K is mapped to itself by degree -3, and after pasting in rigid models we conclude that (K,f) is stably conjugate to $(\mathbf{S}^1, z^{-3})$.

§ 8. Exposed Critical Points

We conclude with the proof of a special case of the conjecture in § 3. Perhaps the argument will shed light on the hard question of the existence of invariant line fields.

The following is a special case of a result of Strebel's [Str, Theorem 7].

THEOREM 8.1 Let U be a finitely connected domain in $\hat{\mathbf{C}}$, and let $\phi : \overline{U} \to \overline{U}$ be a quasiconformal homeomorphism which is the identity on ∂U. Then there is a unique quasiconformal map ψ of minimal dilatation among those isotopic to ϕ rel ∂U. The absolute value of the Beltrami differential $|\psi_{\bar{z}}/\psi_z|$ is constant on U (in fact ψ is a Teichmüller mapping of finite norm.)

DEFINITION: We say $f(z)$ has *exposed critical points* if every critical point lies in Ω or on the boundary of a *component* of Ω.

THEOREM 8.2 If J is connected and f has exposed critical points, then

Mod(J,f) is a finite group.

PROOF: By Proposition 6.10, we may assume that every $[\phi]$ in Mod(J,f) has a unique representative ϕ that commutes with f on the whole of $\hat{\mathbf{C}}$ and is conformal except possibly on J. (Altering f by stable conjugacy preserves the stable modular group.)

Let E be a finite union of components of Ω such that $\overline{E}$ contains all the stable regions, all the critical values of f and $E \subset f^{-1}E$. (E exists by assumption of exposed critical points.) It is easy to check that Mod(J,f) has a subgroup of finite index consisting of $[\phi]$ such that ϕ is the *identity* on E.

Let $V = \hat{\mathbf{C}}$–E. (If V is empty the proof is complete.) Each component of V is a finitely connected plane region. By Strebel's result, ϕ can be isotoped rel ∂V to a unique extremal map ψ which is also the identity on E. Actually V has only finitely many multiply-connected components, so ψ is the identity except on finitely many components of V.

Now let $U = f^{-1}(V)$. Then $f : U \to V$ is a covering map. Since $f \circ \phi = \phi \circ f$ and ψ is isotopic to ϕ, we can form a sequence of lifted maps

$$\psi_0 = \psi, \qquad \psi_{n+1} = f^{-1} \circ \psi_n \circ f$$

where the lift is chosen so each ψ_n is again isotopic to ϕ and the identity on E (compare [M1, 8.1]).

By induction, ψ_n is conformal on $f^{-n}(E)$, and the maximal dilatation of ψ_n is the same as that of ψ (since f is conformal). Consider the component V_0 of V on which ψ has maximal dilatation. For each n, ψ_n has the same dilatation as ψ and is isotopic to ψ rel ∂V, so by Strebel's uniqueness result $\psi = \psi_n$ on V_0. On the other hand, $\cup f^{-n}(E) = \Omega$ is an open dense set, so there is some n such that ψ_n is conformal on an open subset of V_0. Since the dilatation of ψ is constant on V_0, we conclude that ψ is conformal on the whole Riemann sphere (and the identity on E) hence ψ=id. $\qquad\qquad\square$

REMARK: The argument is a generalization of Thurston's theorem on the uniqueness of critically finite maps with given combinatorics [DH2]. There the uniqueness results for Teichmüller mappings of multiply punctured spheres play the same role as the more general result of Strebel's which we use above.

Acknowledgements.

I'd like to thank Hubbard and Sullivan for instruction in rational maps, Marden for guidance in Kleinian groups, and the referee for helpful suggestions.

Department of Mathematics, Northwestern University, Evanston IL 60201

59

REFERENCES

[A]   W. Abikoff. Some remarks on Kleinian groups. In *Advances in the Theory of Riemann Surfaces*, Annals of Math Studies 66 (1971), p.1-6.

[AM]   W. Abikoff, B. Maskit. Geometric decompositions of Kleinian groups. Amer. J. Math. 99 (1977), p.687-698.

[AB]   L. Ahlfors, L. Bers. Riemann's mapping theorem for variable metrics. Annals of Math. 72 (1960), pp. 385-404.

[BR]   L. Bers, H. Royden. Holomorphic families of injections. To appear, Acta Mathematica.

[Bl]   P. Blanchard. Complex analytic dynamics on the Riemann sphere. Bull. AMS 11 (1984), pp.85-141.

[DE]   A. Douady, C. Earle. Conformally natural extensions of homemorphisms of the circle. Acta Mathematica 157 (1986), pp.23-48.

[DH1]   A. Douady, J. Hubbard. On the dynamics of polynomial-like mappings. Ann. sci. Ec. Norm. Sup. 18 (1985), pp.287-344.

[DH2]   A. Douady, J. Hubbard. A proof of Thurston's topological characterization of rational maps. Preprint.

[H]   M. Herman. Exemples de fractions rationelles ayant une orbite dense sur la sphere de Riemann. Bull. Soc.Math. de France 112 (1984), pp.93-142.

[J]   K. Johansson. On the mapping class group of simple 3 manifolds. In *Topology of Low Dimensional Manifolds*, Springer-Verlag Lecture Notes 722 (1979), pp.48-66.

[Ker]   S. Kerckhoff. The Nielsen realization problem. Annals of Mathematics 117 (1983), pp.235-265.

[Kra]   I. Kra. Deformation spaces. In *A Crash Course on Kleinian Groups*, Springer-Verlag Lecture Notes 400 (1974), pp.48-70.

[L]   S. Lattes. Sur l'iteration des substitutions rationelles et les fonctions de Poincare'. CRAS Paris 166 (1918), pp.26-28.

[Ma]   R. Mañe' . Instability of Herman rings. Inv. Math. 81 (1985), pp.459-472.

[MSS]   R. Mañe' , P. Sad, D. Sullivan. On the dynamics of rational maps. Ann. sci. Ec. Norm. Sup. t.16 (1983), pp.193-217.

[Mskt]   B. Maskit. Intersection of component subgroups of Kleinian groups. In *Discontinuous Groups and Riemann Surfaces*, Annals of Math Studies 79 (1974), pp.349-367.

[Mc1]   C. McMullen. Families of rational maps and iterative root-finding algorithms. To appear, Annals of Mathematics.

[Mc2]   C. McMullen. Braiding of the attractor and the failure of iterative algorithms. MSRI Preprint, 1986.

[MB]   J. Morgan, H. Bass (editors). *The Smith Conjecture*. Academic Press (1984).

[Shub]   M. Shub. Expanding maps. In *Global Analysis*, AMS Proc. of Symp. XIV (1970), pp.273-276.

[Str]   K. Strebel. On quasiconformal mappings of open Riemann surfaces. Comm. Math. Helv. 53 (1978), pp.301-321.

[Sul1]   D. Sullivan. Conformal dynamical systems. In *Geometric Dynamics*, Springer-Verlag Lecture Notes 1007 (1983), pp.725-752.

[Sul2]   D. Sullivan. Quasiconformal homeomorphisms and dynamics I: Solution of the Fatou-Julia problem on wandering domains. Annals of Math. 122 (1985), pp.401-418.

[Sul3]   D. Sullivan. Quasiconformal homeomorphisms and dynamics III: Topological conjugacy classes of analytic endomorphisms. Preprint.

[ST]   D. Sullivan, W. Thurston. Extending holomorphic motions. To appear, Acta Mathematica.

[Th]   W. Thurston. *The Geometry and Topology of Three Manifolds*. Lecture notes, Princeton University (1979).

[W]   G. Whyburn. *Analytic Topology*. AMS Coll. Publ. 28 (1942).

[Z]   H. Zieschang. *Finite Groups of Mapping Classes of Surfaces*. Springer-Verlag Lecture Notes 875 (1981).

Selfsimilar zippers

BY KARI ASTALA

Introduction.

If D is a simply connected domain of hyperbolic type in $\overline{\mathbb{C}}$ we denote by ρ_D the density of the hyperbolic metric of D. Then

$$\rho_D(z) = \frac{|g'(z)|}{1 - |g(z)|^2},$$

where $g : D \to \Delta = \{z : |z| < 1\}$ is conformal. The density provides a natural norm

$$(1) \qquad \|\phi\|_D = \sup_{z \in D} |\phi(z)| \rho_D^{-2}(z)$$

for the Schwarzian derivative $S_f = (f''/f')' - \frac{1}{2}(f''/f')^2$ of a locally univalent function f. If f is conformal in a domain D, then $\|S_f\|_D \leqq 12$ and conversely in the unit disk Δ, f is conformal whenever $\|S_f\|_\Delta \leqq 2$ (see [G2], [L] for this and related topics).

Recently Thurston found a new aspect of the Schwarzian derivative and conformal mappings, the existence of *rigid* domains: following [T] we call a domain $D \subset \overline{\mathbb{C}}$ rigid if there exists an $\varepsilon > 0$ such that every conformal map f in D with $\|S_f\|_D < \varepsilon$ is a Möbius transformation.

THEOREM 1.1 (Thurston, [T]). *There exist (uncountably many) rigid domains.*

In this paper we give new proofs for Theorem 1.1. But instead of the incorrigible quasiintervals of Thurston we study only situations that are invariant under suitable similarities or Möbius transformations; in these cases the idea of cascade effect or the interlocking teeth (c.f. [T]) can be more directly applied. We start with a simple proof for the rigidity of the complements of the snowflake curves and then obtain two generalizations. Firstly, the proof is extended to all selfsimilar quasiintervals and secondly, we show that (see [T, 4.4]) if Γ is any finitely generated Möbius group which is quasi-Fuchsian but not Fuchsian, then the complement of every subarc of the limit set $L(\Gamma)$ is rigid.

Finally, the proof of Theorem 1.1 is constructed so that it also yields the related result of Gehring on spirals and the Bers conjecture, Theorem 1.2 below. Equip $S = \{S_f : f$ conformal in $\Delta\}$ with the topology induced by the norm $\| \ \|_\Delta$ in (1) and denote by T the set of those $S_f \in S$ for which f has a quasiconformal extension to $\overline{C}$. Then T is the universal Teichmüller space of Bers and his conjecture was that $S = \overline{T}$, the closure of T. However, the following gave a negative answer.

THEOREM 1.2 (Gehring, [G1]). *Let* $\beta = \{z = \pm e^{(a-i)\log t} : 0 < t \leq 1\} \cup \{0\}$. *If* $0 < a < \frac{1}{8\pi}$, *there exists a constant* $\varepsilon_1 > 0$ *with the following property: if* h *is conformal in* $\overline{C} \setminus \beta$ *and* $\|S_h\|_{\overline{C}\setminus\beta} \leq \varepsilon_1$, *then* $h(\overline{C} \setminus \beta)$ *is not a Jordan domain.*

To see the connection between Theorems 1.1, 1.2 and the geometry of S, recall the transformation rules of the Schwarzian,

$$\|S_f - S_g\|_\Delta = \|S_{f \circ g^{-1}}\|_{g\Delta}.$$

If $g : \Delta \to \overline{C} \setminus \beta$ is conformal and $S_f \in S$ with $\|S_f - S_g\|_\Delta \leq \varepsilon_1$, apply 1.2 to $h = f \circ g^{-1}$. Since $\|S_h\|_{\overline{C}\setminus\beta} \leq \varepsilon_1$, $f(\Delta) = h(\overline{C} \setminus \beta)$ is not a quasidisk. Consequently, no $S_f \in S$ with $\|S_f - S_g\|_\Delta \leq \varepsilon_1$ is contained in T and so $S_g \in S$, $S_g \notin \overline{T}$. In the same manner Theorem 1.1 shows that S is not even connected; if g is conformal and $g(\Delta)$ rigid, S_g is an isolated point of S.

In the sequel our notation will be standard (see [G2] or [L]). In particular, by a quasiinterval we mean the image of the unit interval $[0, 1]$ under a K-quasiconformal map of the finite plane C. A K-quasidisk denotes the image of Δ under a K-quasiconformal map of $\overline{C}$ and a K-quasicircle the boundary of a K-quasidisk. Note, however, that a different notion of rigidity has been used in [G2].

2. Snowflakes.

A quasiinterval α is called *selfsimilar* if there are N similarities σ_i, $1 \leq i \leq N$ and $N \geq 2$, such that

$$(2) \qquad \alpha = \bigcup_{i=1}^{N} \sigma_i(\alpha)$$

and for $i \neq j$ $\sigma_i(\alpha) \cap \sigma_j(\alpha)$ is either empty or consists of a point. The standard example in this connection is the snowflake curve:

EXAMPLE 2.1: Let $a_1 = 0$, $a_2 = 2$, $a_3 = 3 + i\sqrt{3}$, $a_4 = 4$, and $a_5 = 6$ and denote by σ_i the similarity which maps the line segment $\overline{a_1 a_5}$ to $\overline{a_i a_{i+1}}$, $\sigma_i(a_1) = a_i$.

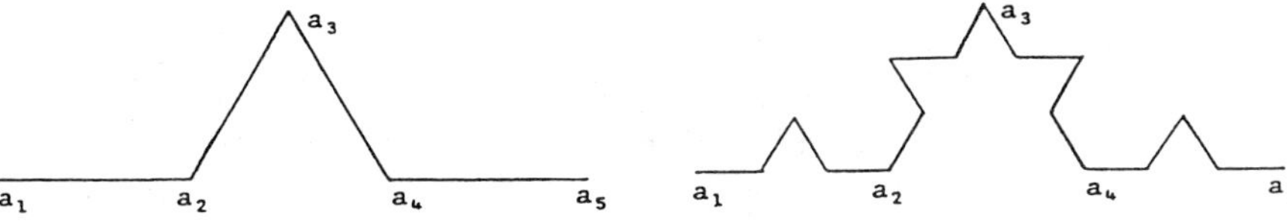

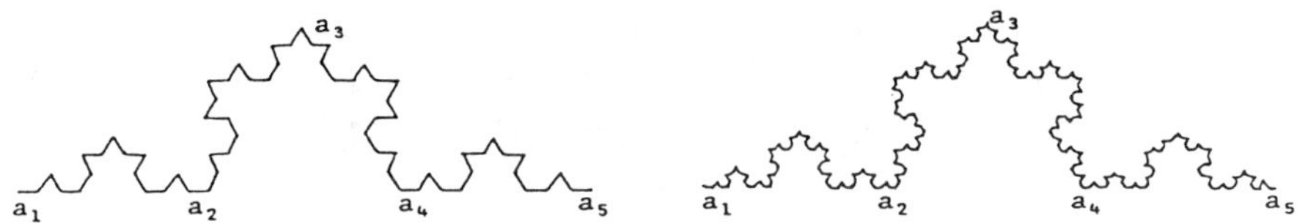

Abbreviating $\Sigma(A) = \cup_1^4 \sigma_i(A)$ there is a unique compact set $A \subset \mathbb{C}$, called the Koch curve or the snowflake curve, such that $\Sigma(A) = A$. In fact the iterated arcs $\Sigma^p(\overline{a_1 a_5}) = \Sigma(\Sigma^{p-1}(\overline{a_1 a_5}))$ approach the snowflake curve in the Hausdorff metric

$$(3) \qquad d_H(A, B) = \max \left\{ \sup_{x \in A} d(x, B), \sup_{y \in B} d(y, A) \right\}$$

when $p \to \infty$ [**H**, section 3]. The above figures show the first four approximations.

In the standard snowflake curve the similarities decrease the length of $\overline{a_1 a_5}$ in the ratio of $\frac{1}{3}$. One can replace the ratio $\frac{1}{3}$ by any number t, $\frac{1}{4} < t < \frac{1}{2}$, and obtain new arcs α_t. We shall also call the α_t's snowflake curves.

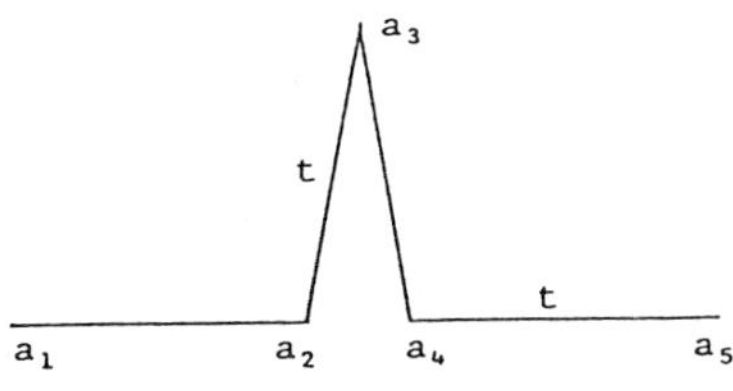

$$|\overline{a_i a_{i+1}}| = t|\overline{a_1 a_5}|$$

From Ahlfors' three point principle ([$\mathbf{A}$] or [$\mathbf{G2}$]) we conclude that snow-flake curves are K-quasiintervals, $K = K(t)$. Furthermore, by [$\mathbf{H}$, 5.3] the Hausdorff dimension of α_t equals $(\log \frac{1}{4})/\log t$. Hence $\dim_H(\alpha_t)$ can have any value in the interval $(1, 2)$.

If α and σ_i are as in (2) we call each point p in $\sigma_i(\alpha) \cap \sigma_j(\alpha)$, $i \neq j$, a *vertex* of α. Suppose p is one of the vertices of α and ψ_k are similarities with $\psi_k(p) = p$ and $\psi_k(z) \to \infty$ as $k \to \infty$, $z \neq p$. Then there exists a subsequence (ψ_j) of (ψ_k) such that $\psi_j(\alpha)$ converges in the Hausdorff metric s_H of $\overline{\mathbb{C}}$ (defined as in (3) with the Euclidean metrid d replaced by the spherical metric s). We denote by $E_p(\alpha)$ the set of all such limits

$$\beta = \lim_{j \to \infty} \psi_j(\alpha).$$

For example, when we translate the vertex $p \in \sigma_i(\alpha_t) \cap \sigma_j(\alpha_t)$ of a snowflake curve to the origin and write $\alpha_{ij} = \sigma_i(\alpha_t) \cup \sigma_j(\alpha_t)$, then $\alpha_{ij} \subset (\frac{1}{t})\alpha_{ij}$ and so

$$(4) \qquad \beta = \bigcup_{n=1}^{\infty} \left(t^{-n}\alpha_{ij}\right) \cup \{\infty\} = \lim_{n \to \infty} t^{-n}\alpha_t \in E_p(\alpha_t).$$

Moreover, in this (translated) case each member $\gamma \in E_p(\alpha_t)$ is a quasicircle with $\gamma = \lambda\beta$, where $\lambda \in \mathbb{C}$ and $t \leq |\lambda| \leq 1$.

The following geometric consequence of (4) represents an infinitesimal version of Theorem 1.1, similar to the one in [$\mathbf{T}$].

LEMMA 2.2. *Let p be a vertex of a snowflake curve $\alpha = \alpha_t$, $\frac{1}{4} < t < \frac{1}{2}$, suppose $\gamma \in E_p(\alpha)$ and let G be one of the components of $\overline{\mathbb{C}} \setminus \gamma$. If ψ is a similarity with $\psi G \subset G$, then $\psi(p) = p$ and $\psi G = G$.*

PROOF: As in (4) we translate p to the origin. Then ψ takes the form $\psi(z) = \lambda z + \delta$ and we must show $\lambda G = G$ and $\delta = 0$.

From (4) we see that $t^n G = G$ for all $n \in \mathbb{Z}$. Hence $\lambda G + \delta \subset G$ implies $\lambda G + t^n \delta \subset G$ which gives $\lambda G \subset \overline{G}$ when $n \to \infty$. Since the map $z \to \lambda z$ is a homeomorphism of $\overline{\mathbb{C}}$, $\lambda G \subset G$.

If $\lambda G \neq G$, there exists a disk $B(z_0, \varepsilon_0) \subset G \setminus \lambda G$. Multiplying z_0 by a power of t we may assume $t \leq |z_0| \leq 1$. Furthermore, as $B(t^n \lambda z_0, t^n |\lambda| \varepsilon_0) \subset \lambda G \setminus \lambda^2 G$, with a suitable n we obtain a disk $B(z_1, \varepsilon_1) \subset \lambda G \setminus \lambda^2 G$ with $t|z_0| \leq |z_1| \leq |z_0|$ and $t\varepsilon_0 \leq \varepsilon_1 \leq \varepsilon_0$. Repeating this argument yields disks of radius ε_i, $t\varepsilon_0 \leq \varepsilon_i \leq \varepsilon_0$, and center z_i, $t^2 \leq |z_i| \leq 1$, contained in

$\lambda^i G \setminus \lambda^{i+1} G$. But since $\lambda^{i+1} G \subset \lambda^i G$, the disks are disjoint and we have a contradiction when $i \to \infty$.

Therefore $\lambda G = G$ and $G + \delta \subset G$. If $\delta \neq 0$, we multiply by t^n in and iterate this inclusion to get $G + kt^n \delta \subset G$ for all $n, k \in \mathbb{N}$. In the limit $G + s\delta \subset \overline{G}$, and hence $G + s\delta \subset G$, for every $s \geqq 0$. In other words, if a line ℓ in the direction $\arg(\delta)$ intersects $\gamma = \partial G$, then $\ell \cap \gamma$ is connected. Every subarc of γ contains an arc similar to α, c.f. (4), which shows that $\ell \cap \gamma$ must consist of a point. However, now γ and also α would be a graph (up to a rotation). This is clearly not the case and so $\delta = 0$. $\qquad\square$

3. Rigidity.

We base our proof of Theorem 1.1 on the well-known extension result of Ahlfors.

THEOREM 3.1 (Ahlfors, [A]). *Let D be a K-quasidisk. Then there exists a constant $\varepsilon_0 = \varepsilon_0(K) > 0$ such that every conformal map f in D with $\|S_f\|_D \leqq \varepsilon_0$ has a quasiconformal extension $F : \overline{\mathbb{C}} \to \overline{\mathbb{C}}$. Moreover we have the dilatation estimate $K(F) \leqq (1 - c\|S_f\|_D)^{-1}$.*

Given a K-quasiinterval α there always exists a K-quasicircle γ passing through ∞ for which $\alpha \subset \gamma$. We denote by G_+ and G_- the components of $\overline{\mathbb{C}} \setminus \gamma$, and if f is conformal in $\overline{\mathbb{C}} \setminus \alpha$, by f_+ and f_- the restrictions of f to these quasidisks. The monotonicity of the hyperbolic metric implies

$$\|S_{f_+}\|_{G_+}, \|S_{f_-}\|_{G_-} \leqq \|S_f\|_{\overline{\mathbb{C}} \setminus \alpha}.$$

Thus Theorem 3.1 shows that if the Schwarzian norm of f is less than $\varepsilon_0(K)$ in $\overline{\mathbb{C}} \setminus \alpha$, f possesses (in G_+ and G_-) one-sided limits at each point $p \in \alpha$. We denote these limits by $f_+(p)$ and $f_-(p)$.

LEMMA 3.2. *Suppose $\alpha = \alpha_t$ is a snowflake curve, $\frac{1}{4} < t < \frac{1}{2}$. There exists a constant $\delta > 0$, depending only on t, such that if*

   i) *$\alpha \subset \beta$, where β is an arc similar to α,*
   ii) *f is conformal in $\overline{\mathbb{C}} \setminus \beta$, f fixes the endpoints of α and*
   iii) *$\|S_f\|_{\overline{\mathbb{C}} \setminus \beta} \leqq \delta$,*

then $f_+(p) = f_-(p)$ for all vertices $p \in \alpha$.

PROOF: Suppose f, β satisfy i), ii), and iii) with $\delta \leqq \varepsilon_0(K(t))$. Then Theorem 3.1 provides quasiconformal extensions F_+, F_- for the restrictions f_+ and f_-. Since f is injective $f_+(G_+)$ and $f_-(G_-)$ do not intersect;

therefore $F = (F_+)^{-1} \circ F_-$ satisfies

$$(5) \qquad FG_- \subset G_-, F(\infty) = \infty \text{ and } F(q_j) = q_j, \quad j = 1, 2,$$

where the q_j are the endpoints of α.

Now, if there is a vertex $p \in \alpha$ for which the claim fails for every positive δ, the above reasoning gives us a sequence $(F_n)_1^\infty$ of quasiconformal maps of $\overline{\mathbb{C}}$ such that $F_n(p) \neq p$, (5) holds for each F_n and, by the dilatation estimate in 3.1, $K(F_n) \to 1$ when $n \to \infty$ (Here also G_- depends on n, but in any case $\alpha \subset \partial G_-$.).

As $F_n(p) \neq p$, there are similarities ϕ_n, η_n fixing p such that $\tilde{F}_n = \phi_n \circ F_n \circ \eta_n^{-1}$ maps p to $p + 1$ and $p - 1$ to p. Clearly

$$(6) \qquad \tilde{F}_n(\eta_n G_-) \subset \phi_n G_-.$$

Next, let $n \to \infty$. With a normal family argument, see for example [**L**, I.2], (5) implies that $F_n(z) \to z$ uniformly in $\overline{\mathbb{C}}$. Consequently, $\phi_n(z)$, $\eta_n(z) \to \infty$, $z \neq p$, and by taking a subsequence we may assume that $\phi_n(\alpha) \to \gamma \in E_p(\alpha)$, $\eta_n(\alpha) \to \hat{\gamma} \in E_p(\alpha)$ and $\tilde{F}_n \to \Phi$. Here Φ is a Möbius transformation with $\Phi(\infty) = \infty$, $\Phi(p) = p + 1$ and $\Phi(p - 1) = p$, i.e. $\Phi(z) = z + 1$.

To conclude the proof we use the special properties of snowflakes described in section 2. Namely, we saw in (4) that $\hat{\gamma} = \psi(\gamma)$ for some similarity ψ with $\psi(p) = p$. Thus if G is the component of $\overline{\mathbb{C}} \setminus \gamma$ which is the limit of the $\phi_n G_-$'s, letting $n \to \infty$ in (6) yields $\Phi(\psi G) = 1 + \psi G \subset G$. This, however, contradicts Lemma 2.2.

In brief, we have shown that for each vertex p there is a constant $\delta(p) > 0$ such that i), ii) and iii) imply $f_+(p) = f_-(p)$; hence we prove the claim when we choose δ to be the smallest of these $\delta(p)$'s. $\qquad \square$

Theorem 1.1 is now an immediate consequence of

THEOREM 3.3. *Let α be a snowflake curve, $1 < \dim_H(\alpha) < 2$. Then $\overline{\mathbb{C}} \setminus \alpha$ is rigid.*

PROOF: The proof applies Lemma 3.2 and the invariance of the Schwarzian; composing f with similarities does not change the norm of S_f.

We assume that f is conformal in $\overline{\mathbb{C}} \setminus \alpha$ and $\|S_f\|_{\overline{\mathbb{C}} \setminus \alpha} \leq \delta$, where δ is the constant of Lemma 3.2. After normalizing f with a similarity so that it fixes the endpoints of α, we see from 3.2 that $f_+(p) = f_-(p)$ for all vertices

p. Hence, if σ_i is any of the similarities in (2), we may apply 3.2 in turn to $f \circ \sigma_i$ and obtain $(f \circ \sigma_i)_+(p) = (f \circ \sigma_i)_-(p)$, that is $f_+(\sigma_i(p)) = f_-(\sigma_i(p))$ for all i and p.

Repeating this process locks the teeth of the snowflake-zipper. In the end $f_+(q) = f_-(q)$ for all q that are images of the vertices p under iterated powers of the σ_i, $1 \leqq i \leqq 4$. Since such points q are dense on α, $f_+(q) = f_-(q)$ for every $q \in \alpha$ and f extends continuously across α.

Because f is conformal in $\overline{\mathbb{C}} \setminus \alpha$ and by 3.1 injective in $\overline{G}_+$, $\overline{G}_-$, f is a homeomorphism of $\overline{\mathbb{C}}$. On the other hand, α is a removable singularity for such f; any homeomorphism of $\overline{\mathbb{C}}$ conformal outside a quasiinterval is conformal in the whole $\overline{\mathbb{C}}$, see [**A**], [**T**, 3.4] or [**L**, I.6.1]. Therefore f is a Möbius transformation whenever $\|S_f\|_{\overline{\mathbb{C}} \setminus \alpha} \leqq \delta$. $\qquad\square$

4. Generalizations.

The above proof of Theorem 1.1 applies also to many other situations such as Theorem 1.2. If β is the logarithmic spiral in 1.2 and $\phi(z) = \tau z$, $\tau = e^{-2\pi a}$, then $\phi(\beta) \subset \beta$, whence $\beta_\infty = \bigcup_1^\infty \phi^{-n}(\beta) \cup \{\infty\} \in E_0(\beta)$ and the reasoning of Lemmas 2.2 and 3.2 works and yields Theorem 1.2. However, to study the limit set of a quasi-Fuchsian group we need the following more technical version. Below the notation is as in section 3, s stands for the spherical distance, $\|S_f\|_{\overline{\mathbb{C}} \setminus \gamma}$ denotes the maximum of $\|S_{f_+}\|_{G_+}$, $\|S_{f_-}\|_{G_-}$ and the limits $f_+(p) = \lim_{z \to p} f_+(z)$, $f_-(p) = \lim_{z \to p} f_-(z)$ are defined as soon as $\|S_f\|_{\overline{\mathbb{C}} \setminus \gamma} \leqq \varepsilon_0(K)$.

THEOREM 4.1. *Let β be a K-quasiinterval which is invariant under a loxodromic Möbius transformation ϕ, $\phi(\beta) \subset \beta$. Let p, q be the fixed points of ϕ and suppose that $p \in \mathrm{int}(\beta)$ and that the condition (7) holds.*

(7)   *In every direction $\theta \in [0, 2\pi]$ there is a circle or a line c_θ through q such that $\beta \cap (c_\theta \setminus \{q\})$ is not connected. (Here θ is determined by the tangent of c_θ at q.)*

Then for every $b > 0$ there exists a constant $\delta > 0$ that has the following property. If the three conditions

    i) *$\beta \subset \gamma$, where γ is a K-quasicircle.*
    ii) *f is conformal in $\overline{\mathbb{C}} \setminus \gamma$ and $\|S_f\|_{\overline{\mathbb{C}} \setminus \gamma} \leqq \delta$.*
    iii) *There are points $x_1, x_2, x_3 \in \gamma$ such that $s(x_i, x_j) \geqq b$ and $f_+(x_i) = f_-(x_i)$, $1 \leqq i \neq j \leqq 3$.*

are satisfied, then $f_+(p) = f_-(p)$.

PROOF: According to our assumptions p is the attractive fixed point of ϕ. Hence we may conjugate β so that $p = 0$, $q = \infty$ and $\phi(z) = \tau z$, $0 < |\tau| < 1$. This does not change $\|S_f\|_{\overline{\mathbb{C}} \setminus \gamma}$ but β need not remain a quasiinterval (if $q \in \beta$) and now $s(x_i, x_j) \geq b' > 0$ where b' depends only on b and the conjugation map.

If the claim in 4.1 is not true, we obtain as in Lemma 3.2 a sequence of K_n-quasiconformal maps $F_n : \overline{\mathbb{C}} \to \overline{\mathbb{C}}$ and a sequence of quasidisks G_n such that $\beta \subset \gamma_n = \partial G_n$, $K_n \to 1$ when $n \to \infty$, $F_n G_n \subset G_n$, F_n fixes three points x_n^1, x_n^2, x_n^3 with $s(x_n^i, x_n^j) \geq b'$ and $F_n(0) \neq 0$. The last condition shows that for some complex numbers λ_n, μ_n the map $\tilde{F}_n(z) = \lambda_n F_n(\mu_n^{-1} z)$ takes 0 to 1 and -1 to 0.

Since F_n fixes the points x_n^i, $F_n \to$ id uniformly on $\overline{\mathbb{C}}$. Therefore $\lambda_n, \mu_n \to \infty$ and for at least one index i we have $\lambda_n x_n^i \to \infty$, $\mu_n x_n^i \to \infty$. As $\tilde{F}_n(\mu_n x_n^i) = \lambda_n x_n^i$, the $\tilde{F}_n$ form a compact sequence of homeomorphisms of $\overline{\mathbb{C}}$. In the limit $\tilde{F}_n \to \psi$, $\psi(z) = z + 1$, uniformly on $\overline{\mathbb{C}}$.

Because $\phi(\beta) \subset \beta$ and $0 \in \text{int}(\beta)$, $\beta_\infty = \bigcup_1^\infty \phi^{-n}(\beta) \cup \{\infty\} \in E_0(\beta)$. Then each $\alpha \in E_0(\beta)$ is a multiple of β_∞ and we can take subsequences, also denoted by (λ_n) and (μ_n), such that $\lambda_n \beta \to a_1 \beta_\infty$ and $\mu_n \beta \to a_2 \beta_\infty$. In addition, we may assume that $\lambda_n G_n \to a_1 G$ and $\mu_n G_n \to a_2 G$, where G is a component of $\overline{\mathbb{C}} \setminus \beta_\infty$. As $\tilde{F}_n(\mu_n G_n) \subset \lambda_n G_n$ we finally obtain $a_2 G + 1 \subset a_1 G$.

However, like in the proof of Lemma 2.2, the inclusion $a_2 G + 1 \subset a_1 G$ (combined with the invariance of β and β_∞ under ϕ) yields first that $a_2 G = a_1 G$ and then that the intersection of β_∞ with any line ℓ in the direction $\arg(a_1)$ is connected. As β is a subarc of the quasicircle β_∞, we get a contradiction with (7). This proves the claim. $\square$

EXAMPLES 4.2: By Theorem 4.1, if β is any of the curves below, then $S_g \in S \setminus \overline{T}$ for each conformal $g : \Delta \to \overline{\mathbb{C}} \setminus \beta$. (c.f. section 1; $f(\overline{\mathbb{C}} \setminus \beta)$ cannot be a Jordan domain if $f_+(p) = f_-(p)$ for some $p \in \text{int}(\beta)$.) In particular, the first example gives again Theorem 1.2.

a) $\beta = \left\{ z = \pm e^{(a-i)\log t} : 0 < t \leqq 1 \right\} \cup \{0\}, \quad 0 < a < \infty, \quad \phi(z) = e^{-2\pi a} z$

b)

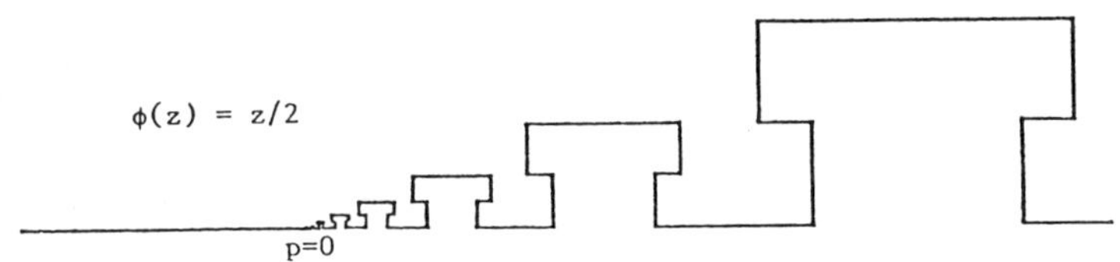

c)

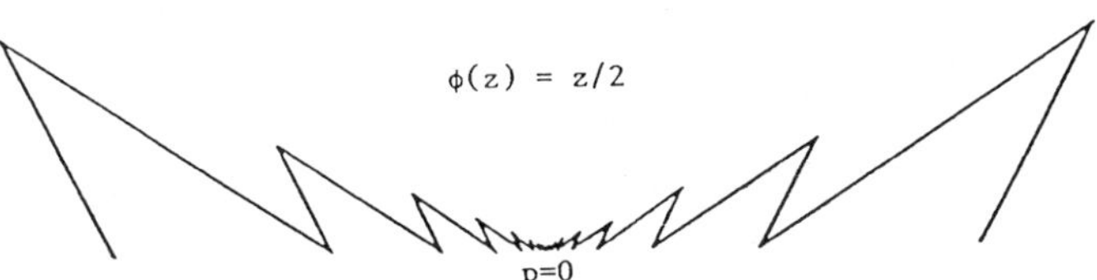

In Theorem 4.1 the condition (7) can be dropped if ϕ^2 is strictly loxodromic.

LEMMA 4.3. *Let β be a Jordan arc with $0 \in \operatorname{int}(\beta)$, suppose $\tau\beta \subset \beta$, $0 < |\tau| < 1$, and let ℓ be a line through the origin.*

   a) *If $0 < |\arg\tau| < \pi$, then $\ell \cap \beta$ is not connected.*
   b) *If $\arg\tau = \pi$ and if there is a Jordan domain G such that $\beta \subset \partial G$ and $\tau G = G$, then $\ell \cap \beta$ is not connected.*

PROOF: a)   Replacing τ by τ^k whenever necessary, we may assume that $0 < \arg\tau \le \frac{2\pi}{3}$. If ℓ intersects β only at the origin, so do $\tau^{-1}\ell$, $\tau^{-2}\ell$ and so on. These lines $\tau^{-n}\ell$ divide the plane into disjoint open sectors and now there are at least three sectors containing points of β. This argument proves that $\beta \cap \ell$ contains nonzero points $z \in \mathbb{C}$. As $\tau\beta \subset \beta$, β is not a line segment and so $\ell \cap \beta$ is not connected.

   b)   We can assume that there is a point $z \in \ell \cap G$ since ∂G cannot be a line. Choose a path γ connecting z to τz in G. If the segment $I \subset \ell$ between z and $\tau^2 z$ does not intersect ∂G, then $\gamma \cup \tau\gamma \cup I$ surrounds $0 \in \partial G$ in G and this is impossible. Replacing z by $\tau^{2k}z$ we see that I intersects β and that $\beta \cap \ell$ is not connected. $\square$

Suppose that Γ is a group of Möbius transformations in $\overline{\mathbb{C}}$. We here call Γ *quasi-Fuchsian* if the limit set $L(\Gamma)$ is a quasicircle and *Fuchsian* if $L(\Gamma)$ is a circle or a line.

COROLLARY 4.4. *Let Γ be a finitely generated Möbius group which is quasi-Fuchsian but not Fuchsian. If $\beta \subset L(\Gamma)$ is a closed subarc, $\beta \ne L(\Gamma)$, then $\overline{\mathbb{C}} \setminus \beta$ is rigid.*

PROOF: Let $\gamma = L(\Gamma)$ be the quasicircle in the assumption i) of 4.1 and suppose G is a component of $\overline{\mathbb{C}} \setminus \gamma$. As Γ and the stabilizer Γ_G have the same loxodromic fixed points, we can assume that Γ preserves G. Furthermore, as Γ is not Fuchsian it contains a strictly loxodromic element g [B, Theorem 5.2.1]. If p, q are the fixed points of g and if $p \in \operatorname{int}(\beta)$ and $g(\beta) \subset \beta$, Lemma 4.3 gives (7).

Choose three distinct strictly loxodromic fixed points $p_1, p_2, p_3 \in \operatorname{int}(\beta)$ and replace β by a subarc β_k so that all assumptions of Theorem 4.1 hold for each p_k, $k = 1, 2, 3$. Choose also three points $x_1, x_2, x_3 \in L(\Gamma) \setminus \beta$ and, if $g_1, \ldots, g_n$ is a set of generators of Γ, a constant $b > 0$ so that $s(x, y) \ge b$

for all $x, y \in X$, $x \neq y$, where

$$X = \left\{ x_k, g_i(p_k), g_i^{-1}(p_k) : 1 \leq i \leq n, 1 \leq k \leq 3 \right\}.$$

Corresponding to this b and to the points p_1, p_2, p_3 there exists a $\delta > 0$ such that $f_+(p_k) = f_-(p_k)$, $k = 1, 2, 3$, whenever f satisfies the hypothesis ii) and iii) of 4.1.

Next let f be a conformal map in $\overline{\mathbb{C}} \setminus \beta$ with $\|S_f\|_{\overline{\mathbb{C}} \setminus \beta} \leq \delta$. As $f_+(x_k) = f_-(x_k) = f(x_k)$, Theorem 4.1 shows that $f_+(z) = f_-(z)$ for $z = p_1, p_2, p_3$. Furthermore, these equalities are inherited to the orbits $\Gamma(p_k)$: Denote by Γ_m the set of elements $g \in \Gamma$ which are of the form $g = g_{i_1}^{\pm 1} \circ \cdots \circ g_{i_m}^{\pm 1}$, $m \in \mathbb{N}$, and note that $\|S_{f \circ g}\|_{\overline{\mathbb{C}} \setminus \gamma} = \|S_f\|_{\overline{\mathbb{C}} \setminus \gamma}$ for all $g \in \Gamma$. If $g \in \Gamma_1$,

$$(8) \qquad (f \circ g)_+(z) = (f \circ g)_-(z)$$

when $z = g^{-1}(p_k) \in X$, $k = 1, 2, 3$. So according to 4.1, (8) is satisfied by $z = p_1$, $z = p_2$ and $z = p_3$, too.

Suppose then that $g \in \Gamma_{m+1}$ and that $f_+(y) = f_-(y)$ whenever $y = h(p_k)$, $h \in \Gamma_m$ and $1 \leq k \leq 3$. As $g = h \circ g_i^{\pm 1}$, where $h \in \Gamma_m$, (8) holds for $z = g_i^{\mp 1}(p_k) \in X$, $k = 1, 2, 3$. Thus 4.1 applies and we see that $f_+(y) = f_-(y)$ if $y = g(p_k)$ and $g \in \Gamma_{m+1}$.

This inductive argument, based again on a kind of cascade effect, proves that (8) is true for all z in the orbit $\Gamma(p_1)$. In particular, $f_+(y) = f_-(y)$ if $y \in \beta$, i.e. f extends continuously across β. As in the proof of Theorem 3.3 f must then be a Möbius transformation. $\qquad \square$

Similar ideas work also for all selfsimilar quasiintervals. Note, however, that such curves are always incorrigible in the sense of Thurston and so our last result is a special case of $[\mathbf{T}, \text{Theorem } 4.3]$.

COROLLARY 4.5. *Let β be a selfsimilar quasiinterval which is not a line segment. Then $\overline{\mathbb{C}} \setminus \beta$ is rigid.*

PROOF: Represent β by the similarities σ_i, $1 \leq i \leq N$, as in (2) and let ℓ be the line through the endpoints w_1, w_2 of β. If N or $N - 1$ of the σ_i map ℓ onto a parallel line, β would have to be a line segment. Hence we can assume that σ_1^2, σ_2^2 are strictly loxodromic. If necessary, replace σ_1 or σ_2 by $(\sigma_j)^k \circ \sigma_1$ or $(\sigma_j)^k \circ \sigma_2$ so that their finite fixed points $p_1, p_2 \in \text{int}(\beta)$.

As in 4.4 write

$$X = \{\infty, w_1, w_2\} \cup \left\{ \sigma_i^{-1}(p_1), \sigma_i^{-1}(p_2) : 1 \leq i \leq N \right\},$$
$$b = \min\{s(x, y) : x, y \in X, x \neq y\} > 0$$

and denote by Σ_n the set of similarities of the form $\sigma = \sigma_{i_1} \circ \cdots \circ \sigma_{i_n}$, $\Sigma_0 = \{\mathrm{id}\}$. Let $\delta > 0$ be the constant (corresponding to b, β and the points $p = p_1, p_2$) given by Theorem 4.1 and Lemma 4.3.

If f is conformal in $\overline{\mathbb{C}} \setminus \beta$ with $\|S_f\|_{\overline{\mathbb{C}} \setminus \beta} \leq \delta$, apply Theorem 4.1 first to f and $\gamma = \gamma_k \equiv \bigcup_1^\infty (\sigma_k)^{-n}(\beta) \cup \{\infty\}$, $k = 1, 2$. Then

$$(f \circ \sigma)_+(z) = (f \circ \sigma)_-(z), \quad z = p_1, p_2$$

holds for $\sigma \in \Sigma_0$. And if it holds for all $\sigma \in \Sigma_n$, let $\hat{\sigma} = \sigma \circ \sigma_i$, where $\sigma \in \Sigma_n$, and now apply 4.1 to $f \circ \hat{\sigma}$, to $\gamma = (\hat{\sigma})^{-1}(\gamma_k) \supset \beta$ and to $x_1 = \infty$, $x_2 = \sigma_i^{-1}(p_1)$, $x_3 = \sigma_i^{-1}(p_2)$. In this manner we obtain $f_+(z) = f_-(z)$ for $z = \sigma_{i_1} \circ \cdots \circ \sigma_{i_n}(p_1)$, $n \in \mathbb{N}$. Such points z are dense on β [**H**, 3.1], thus f extends across β and, consequently, f is a Möbius transformation, c.f. 3.3.

$\square$

Department of Mathematics, University of Helsinki, Helsinki, FINLAND

References

[**A**] Ahlfors, L.V., *Quasiconformal reflections*, Acta Math. **109** (1963), 291–301.

[**B**] Beardon, A.F., "The Geometry of Discrete Groups," Springer Verlag, 1983.

[**G1**] Gehring, F.W., *Spirals and the universal Teichmüller space*, Acta Math. **141** (1978), 99–113.

[**G2**] Gehring, F.W., "Characteristic Properties of Quasidisks," Les Presses de l'Université de Montréal, 1982.

[**H**] Hutchinson, J.E., *Fractals and selfsimilarity*, Indiana Univ. Math. J. **30** (1981), 713–747.

[**L**] Lehto, O., "Univalent functions and Teichmüller spaces," Springer Verlag, 1986.

[**T**] Thurston, W.P., *Zippers and univalent functions*, in "The Bieberbach Conjecture," Math Surveys No. 21, Amer. Math. Soc., Providence, 1986, pp. 185–197.

On the boundary behavior of
Bloch functions

BY J.L. FERNÁNDEZ AND CH. POMMERENKE

The function f is called a Bloch function if it is analytic in the unit disk $\mathbb{D}$ and satisfies one of the following equivalent [8] conditions:

(i) $(1 - |z|^2)|f'(z)|$ is bounded in $\mathbb{D}$;

(ii) the Riemann image surface over $\mathbb{C}$ contains no arbitrarily large schlicht disks;

(iii) $f = c \log g'$ where g is univalent in $\mathbb{D}$ and c is a suitable constant.

The Hausdorff dimension $\dim E$ is the infimum of all α such that, for every $\varepsilon > 0$, there are countably many disks D_k with

$$E \subset \bigcup_k D_k, \qquad \sum_k (\text{diam } D_k)^\alpha < \varepsilon.$$

If $E \subset \partial\mathbb{D}$ then $0 \le \dim E \le 1$ and

$$\dim E < 1 \Rightarrow \text{mes } E = 0, \qquad \dim E > 0 \Rightarrow \text{cap } E > 0;$$

neither implication can be reversed.

Let $A = A(f)$ denote the set of all $\varsigma \in \partial\mathbb{D}$ for which either

$$(1) \qquad \lim_{r \to 1} f(r\varsigma) \ne \infty \quad \text{or} \quad \lim_{r \to 1} Re f(r\varsigma) = +\infty$$

exists. If f is a Bloch function then these radial limits can be replaced by angular limits or, on the other hand, by asymptotic values [7] [10, Section 9.1]. The function g defined by $g' = e^{-f}$ has a finite angular derivative at $\varsigma \in \partial\mathbb{D}$ if and only if $\varsigma \in A$. The set A is uncountably dense on $\partial\mathbb{D}$ by a result of R.L. Hall [3].

The Hadamard gap series

$$f(z) = \sum_{k=0}^{\infty} b_k z^{2^k}, \qquad 0 < \limsup_{k \to \infty} |b_k| < \infty$$

is a Bloch function without any finite angular limits. It satisfies $\dim A > 0$ [2] [5] and even $\dim A = 1$ if for instance $b_k = 1$, as J. Hawkes [4] has shown.

PROBLEM 1: Does $\dim A = 1$ hold for every Bloch function?

It is not even known whether $\dim A > 0$ is always true. This would give a positive answer to the question [9, Problem 3.4] whether cap $A > 0$ is true.

Let $B = B(f)$ denote the set of all $\varsigma \in \partial D$ such that

$$\text{(2)} \qquad \limsup_{r \to 1} |f(r\varsigma)| < \infty.$$

If f is a Bloch function then (2) holds if f is bounded on a curve ending at ς [1, Th. 4.2] [10, Th. 9.5]. It is only known [6] that B is uncountably dense on ∂D.

PROBLEM 2: Does $\dim B = 1$ hold for every Bloch function?

We can prove this only for a very special case by using a result of D. Sullivan [11] on the limit sets of Fuchsian groups.

THEOREM. *Let G be a domain in $\mathbb{C}$ that contains no arbitrarily large disks. If f is a universal covering map of D onto G then $\dim B = 1$.*

By (ii), our assumption about G holds if and only if f is a Bloch function.
PROOF: Since f is a universal covering map there is a Fuchsian group Γ such that

$$\text{(3)} \qquad f(z') = f(z) \iff \exists \gamma \in \Gamma : z' = \gamma(z).$$

Let $L_0(\Gamma)$ denote the angular limit set, i.e. the set of all $\varsigma \in \partial D$ such that there are infinitely many points $\gamma(0)$ ($\gamma \in \Gamma$) in some Stolz angle at ς.

If cap $\partial G > 0$ then f has a finite angular limit almost everywhere on ∂D, so that mes $B > 0$. Hence we may assume cap $\partial G = 0$. It follows [12, Th. XI.20] that mes $L_0(\Gamma) = 2\pi$.

Let $\Gamma = \{\gamma_\mu : \mu = 1, 2, \ldots\}$ and let Γ_n be the subgroup generated by $\gamma_1, \ldots, \gamma_n$. We choose $r_n < 1$ such that $|\gamma_\mu(0)| < r_n$ for $\mu = 1, \ldots, n$. Since f is bounded on $D_n = \{|z| < r_n\}$ it follows from (3) that f is bounded on

$$\text{(4)} \qquad V_n = \bigcup_{\gamma \in \Gamma_n} \gamma(D_n).$$

Since $\gamma_\mu(0) \in D_n$ for all generators γ_μ of Γ_n we conclude that V_n is connected. Since f is a Bloch function it follows that (2) holds for every $\varsigma \in \partial V_n \cap \partial D$. Thus, by (4),

$$\text{(5)} \qquad L_0(\Gamma_n) \subset \partial V_n \cap \partial D \subset B \qquad \text{for all } n.$$

Since (Γ_n) increases to Γ, a result of Sullivan [**11**, Cor. 6, Cor. 27] shows that

$$\dim L_0(\Gamma_n) \to \dim L_0(\Gamma) = 1 \qquad \text{as } n \to \infty$$

because mes $L_0(\Gamma) > 0$. Hence it follows from (5) that $\dim B = 1$.

J.L. Fernández, Dept. of Math., University of Maryland, College Park, MD 20742

Ch. Pommerenke, Fachbereich Mathematik, Technische Universität, D-1000 Berlin 12, Germany

References

1. Anderson, J.M., Clunie, J., and Pommerenke, Ch., *On Bloch functions and normal functions*, J. Reine Angew. Math. **270** (1974), 12–37.
2. Csordas, G., Lohwater, A.J., and Ramsey, T., *Lacunary series and the boundary behavior of Bloch functions*, Michigan Math. J. **29** (1982), 281–288.
3. Hall, R.L., *On the asymptotic behaviour of functions holomorphic in the unit disk*, Math. Z. **107** (1968), 357–362.
4. Hawkes, J., *Probabilistic behaviour of some lacunary series*, Z. Wahrsch. Verw. Geb. **53** (1980), 21–33.
5. Gnuschke, D., and Pommerenke, Ch., *On the radial limits of functions with Hadamard gaps*, Michigan Math. J. **32** (1985), 21–31.
6. Gnuschke-Hauschild, D., and Pommerenke, Ch., *On Bloch functions and gap series*, J. Reine Angew. Math. **367** (1986), 172–186.
7. Lehto, O., and Virtanen, K.I., *Boundary behaviour and normal meromorphic functions*, Acta Math. **97** (1957), 47–65.
8. Pommerenke, Ch., *On Bloch functions*, J. London Math. Soc. (2) **2** (1970), 689–695.
9. Pommerenke, Ch. (ed.), *Probleme aus der Funktionentheorie*, Jahresber. Deutsch. Math.-Ver. **73** (1971), 1–5.
10. Pommerenke, Ch., "Univalent Functions," Vandenhoeck & Ruprecht, Göttingen, 1975.
11. Sullivan, D., *The density at infinity of a discrete group of hyperbolic motions*, Publ. IHES **50** (1979), 419–450.
12. Tsuji, M., "Potential Theory in Modern Function Theory," Maruzen, Tokyo, 1959.

Harmonic functions in quasicircle domains

BY A. HINKKANEN

Abstract. Let G be a bounded plane domain whose boundary consists of finitely many disjoint quasicircles. Let u be a harmonic in G and continuous in $\bar{G}$, the closure of G. We assume that the modulus of continuity of u is majorized by a suitable type of function on the boundary of G and obtain estimates for the modulus of continuity of u in $\bar{G}$. We show by examples that in a certain sense these estimates are best possible.

1. Introduction and Results.

1.1. Let G be a bounded plane domain. Let u be a real-valued function harmonic in G and continuous in the closure $\bar{G}$ of G. If the nonnegative nondecreasing continuous function $\mu(t)$ defined for $t \geq 0$ satisfies

$$(1.1) \qquad \mu(2t) \leq 2\mu(t)$$

for $t \geq 0$, we call μ a *majorant*. We assume that the modulus of continuity of u on ∂G, the boundary of G, is majorized by μ, that is,

$$(1.2) \qquad |u(z_1) - u(z_2)| \leq \mu(|z_1 - z_2|)$$

for all $z_1, z_2 \in \partial G$ and ask what kind of upper bound can be found for $|u(z_1) - u(z_2)|$ when $z_1, z_2 \in \bar{G}$.

In [6, Theorem 1] we obtained the following result, where $\log^+ x = \max(0, \log x)$ and where diam G denotes the diameter of G.

THEOREM A. *Let G, u and μ be as above, and suppose that (1.2) holds for all $z_1, z_2 \in \partial G$. Suppose further that there is a positive number d such that every component of ∂G, other than an isolated point, has diameter not less than d. Then*

$$(1.3) \qquad \begin{aligned} |u(z_1) - u(z_2)| \leq A\mu(|z_1 - z_2|)(\text{diam } G)(d^{-1} + (d|z_1 - z_2|)^{-1/2})/ \\ \log(2d^{-1} \text{ diam } G) \end{aligned}$$

for all $z_1, z_2 \in \bar{G}$ where A is an absolute constant.

This work was supported in part by the U.S. National Science Foundation. AMS (1980) Classification. Primary 30C60.

If $\mu(t) = Mt^\alpha$ where $M > 0$ and $0 < \alpha < 1$, we can replace (1.3) by

$$(1.4) \qquad |u(z_1) - u(z_2)| \leq C_\alpha \mu(|z_1 - z_2|)(\operatorname{diam} G)^\alpha (d^{-\alpha} + d^{-1/2}|z_1 - z_2|^{\frac{1}{2}-\alpha})$$
$$/ \log(2d^{-1} \operatorname{diam} G)$$

if $\frac{1}{2} < \alpha < 1$, by

$$(1.5) \qquad |u(z_1) - u(z_2)| \leq A_1 \mu(|z_1 - z_2|)\left(\sqrt{\frac{\operatorname{diam} G}{d}}\,/ \log(2d^{-1} \operatorname{diam} G) + \log^+ \frac{d}{|z_1 - z_2|}\right)$$

if $\alpha = \frac{1}{2}$, and by

$$(1.6) \qquad |u(z_1) - u(z_2)| \leq C_\alpha \mu(|z_1 - z_2|)(1 + d^{-\alpha}(\operatorname{diam} G)^\alpha)/ \log(2d^{-1} \operatorname{diam} G)$$

if $0 < \alpha < \frac{1}{2}$, where A_1 is an absolute constant and C_α depends on α only.

In particular, if we denote by α and β the exponents with respect to which u is Hölder continuous on ∂G and in $\bar{G}$, respectively, then Theorem A states that

$$\alpha > \frac{1}{2} \text{ implies that } \beta \geq \frac{1}{2}$$

and that

$$0 < \alpha < \frac{1}{2} \text{ implies that } \beta = \alpha.$$

In [6, Theorem 2] it is shown that at least as far as exponents of Hölder continuity are concerned, Theorem A is, in general, best possible. So $\alpha_0 = \frac{1}{2}$ is, in general, the largest possible number such that $\beta = \alpha$ provided that $0 < \alpha < \alpha_0$.

We remark that in all results like this, if (1.2) holds for a fixed $z_1 \in \partial G$ and all $z_2 \in \partial G$, then (1.3) (and one of (1.4)-(1.6), if relevant) holds for this z_1 and all $z_2 \in \bar{G}$. The method of proof deals with each such fixed z_1 separately.

If ∂G consists of finitely many disjoint C^2-Jordan curves, then by [6, Theorem 3], Theorem A can be sharpened, and in particular the exponent α_0 is equal to 1. In this case one obtains essentially the same results as when G is the unit disk.

Further if $z_1, z_2 \in G$ and $|z_1 - z_2| \leq d(z_1, \partial G)/2$, for instance, where $d(z, \partial G)$ is the distance of z from ∂G, then (1.3)-(1.6) can be improved as observed in [6, Corollary 1].

1.2. We recall that a *K-quasidisk* is the image of the unit disk under a K-quasiconformal mapping of the whole plane. M. Zinsmeister [**8**, Theorem 2] has shown that if G is a K-quasidisk with $1 \leq K \leq \sqrt{2}$, then the exponent α_0 can be improved to $\alpha_0 = 1/K^2$. More precisely, if there is a K-quasiconformal mapping of the whole plane that takes the unit disk *conformally* onto G, one can take $\alpha_0 = 1/K$, which improves $\alpha_0 = 1/2$ when $1 \leq K < 2$. The purpose of this note is to point out that $\alpha_0 = 1/2$ can be improved for *any* bounded quasidisk G, and that the same is true also for finitely connected domains bounded by quasicircles. As we use a different characterization of quasicircles, we cannot say if our value of α_0 improves the bound $1/K$ for all $K \geq 1$, even though this must be so for K sufficiently close to 2.

1.3. We use the following *cross ratio* definition of a quasicircle, which is due to Gehring and seems to have been first used by Blevins [**2,3**]. It seems to be motivated by the geometric characterization of quasicircles due to Ahlfors ([**1**, Theorem 1, p. 295] and the remark after it).

DEFINITION. *Let* Γ *be a Jordan curve in the sphere, and suppose that* $0 < k \leq 1$. *We say that* Γ *is a k-quasicircle if*

$$(1.7) \qquad \frac{|z_1 - z_2||z_3 - z_4| + |z_2 - z_3||z_1 - z_4|}{|z_2 - z_4||z_1 - z_3|} \leq \frac{1}{k}$$

for each quadruple z_1, z_2, z_3, z_4 *of distinct points of* Γ *that follow each other in the positive or negative direction on* Γ.

If some z_i is ∞, then (1.7) is interpreted in the obvious way.

We note that a 1-quasicircle is a circle or line.

If G is a bounded plane domain whose boundary consists of finitely many disjoint k-quasicircles, we say that G is a *k-quasicircle domain*. If G is a k-quasicircle domain for some $k \in (0,1]$, we say that G is a *quasicircle domain*. We could also allow G to have finitely many isolated boundary points since these would be removable singularities for harmonic functions continuous in $\bar{G}$. For the sake of simplicity we omit this possibility that would only lead to a trivial extension of our results.

1.4. When G is a k-quasicircle domain, we write

$$(1.8) \qquad \kappa = (\pi/2)(\pi - \arcsin k)^{-1} \in (1/2, 1]$$

and $d = (1/6)\min(d_1, d_2)$ where d_1 is the shortest diameter of the components of ∂G and d_2 is the shortest distance between any two distinct components of ∂G. So $d = (\text{diam } G)/6$ if G is simply connected.

We prove the following result.

THEOREM 1. *Let G be a k-quasicircle domain, let u be harmonic in G and continuous in $\bar{G}$, and suppose that (1.2) holds for all $z_1, z_2 \in \partial G$ and some majorant μ. Then we have for all $z_1, z_2 \in \bar{G}$ that*

$$(1.9) \qquad |u(z_1) - u(z_2)| \leq C_1 \mu(|z_1 - z_2|) \left(\frac{\text{diam } G}{d} + \frac{\text{diam } G}{d^\kappa |z_1 - z_2|^{1-\kappa}} \right)$$

if $0 < k < 1$ where C_1 depends on $1 - \kappa$ only, and

$$(1.10) \qquad |u(z_1) - u(z_2)| \leq A_1 \mu(|z_1 - z_2|) \left(\frac{\text{diam } G}{d} + \log \frac{\text{diam } G}{|z_1 - z_2|} \right)$$

if $k = 1$ where A_1 is an absolute constant.

If $\mu(t) = Mt^\alpha$ where $M > 0$ and $0 < \alpha \leq 1$, we can replace (1.9) and (1.10) by

$$(1.11) \quad |u(z_1) - u(z_2)| \leq C_2 \mu(|z_1 - z_2|) \left\{ \left(\frac{\text{diam } G}{d} \right)^\alpha + \frac{(\text{diam } G)^\alpha}{d^\kappa |z_1 - z_2|^{\alpha - \kappa}} \right\}$$

if $\kappa < \alpha \leq 1$, by

$$(1.12) \qquad |u(z_1) - u(z_2)| \leq A_2 \mu(|z_1 - z_2|) \left\{ \left(\frac{\text{diam } G}{d} \right)^\kappa + \log \frac{\text{diam } G}{|z_1 - z_2|} \right\}$$

if $\alpha = \kappa$, and by

$$(1.13) \qquad |u(z_1) - u(z_2)| \leq C_3 \mu(|z_1 - z_2|)(1 + d^{-\alpha}(\text{diam } G)^\alpha)$$

if $0 < \alpha < \kappa$. Here C_2 and C_3 depend only on $|\alpha - \kappa|$ while A_2 is an absolute constant.

If (1.2) holds for a fixed $z_1 \in \partial G$ and all $z_2 \in \partial G$, then the relevant inequalities among (1.9)-(1.13) are valid for this z_1 and all $z_2 \in \bar{G}$.

We can take, for example, $A_1 = A_2 = 4 \cdot 10^6$, $C_1 = 4 \cdot 10^6 (1 - \kappa)^{-1}$, $C_2 = 2 \cdot 10^6 (\alpha - \kappa)^{-1}$, and $C_3 = 3 \cdot 10^6 (\kappa - \alpha)^{-1}$.

In particular, if u is Hölder continuous on ∂G with exponent α where $0 < \alpha < \kappa$, then u is Hölder continuous in $\bar{G}$ with the same exponent α. So the exponent α_0 satisfies $\alpha_0 \geq \kappa$, and indeed we may have $\alpha_0 = \kappa$ as the following result shows.

THEOREM 2. *For any $k \in (0,1]$ there is a bounded quasidisk G whose boundary is a rectifiable k-quasicircle, with the following properties. If $\kappa \leq \alpha \leq 1$, there is a function u harmonic in G and continuous in $\bar{G}$ satisfying (1.2) for all $z_1, z_2 \in \partial G$ with $\mu(t) = t^\alpha$, but there is a point $w \in \partial G$ and a sequence z_n of points of G tending to w as $n \to \infty$ such that for all large n*

$$(1.14) \qquad |u(z_n) - u(w)| \geq A\mu(|z_n - w|)|z_n - w|^{\kappa - \alpha}(\alpha - \kappa)^{-1}$$

if $\kappa < \alpha \leq 1$ and

$$(1.15) \qquad |u(z_n) - u(w)| \geq A\mu(|z_n - w|) \log(|z_n - w|^{-1})$$

if $\alpha = \kappa$ where A is a positive absolute constant.

The lower bound in (1.14) also shows that the constants C_1 and C_2 in (1.9) and (1.11) cannot be essentially improved as regards their dependence on $1 - \kappa$ and $\alpha - \kappa$.

1.5. The inequality (1.13) has the following consequences.

COROLLARY 1. *If Γ is a bounded k-quasicircle, and if f is Hölder continuous on Γ with exponent $\alpha < \kappa$, then f can be extended to a function F in the plane that is Hölder continuous with exponent α.*

Here f can be real or complex valued, and one can take F to be harmonic outside Γ. Zinsmeister observed that [8, Theorem 2] has a similar consequence [8, Proposition 2], and the proof of Corollary 1 is the same. Therefore we only mention that in the interior domain G of Γ, the function F is taken to be the harmonic extension of f to G, and that if $0 \in G$, for example, then for z in the complement $\mathbb{C} \backslash \bar{G}$ of $\bar{G}$, the function $F(1/z)$ is taken to be the harmonic extension of $f(1/z)$ from Γ^{-1} to $(\mathbb{C} \backslash \bar{G})^{-1}$ where $E^{-1} = \{1/z : z \in E\}$.

COROLLARY 2. *Let G be a bounded quasidisk, let ∂G be a k-quasicircle, and let u be harmonic in G and continuous in $\bar{G}$. If u is Hölder continuous on ∂G with exponent $\alpha < \kappa$, then u and its harmonic conjugate are Hölder continuous in $\bar{G}$ with exponent α.*

As Zinsmeister [8] points out, this kind of result follows from (1.13), the fact that $|\nabla u| = |\nabla v|$ in G where v is the harmonic conjugate of u, and the following inequality of Gehring and Martio [**4**, Theorem 1.1, p. 68 and Theorem 2.1, p. 70]. We have

$$\alpha c \|u\|_\alpha \le \sup_{z \in G} d(z, \partial G)^{1-\alpha} |\nabla u(z)| \le (4/\pi) \|u\|_\alpha$$

where $\|u\|_\alpha$ is the smallest number M such that

$$|u(z_1) - u(z_2)| \le M |z_1 - z_2|^\alpha \text{ for all } z_1, z_2 \in G$$

and the positive constant c depends on k only.

If $\alpha = \kappa$ in Corollary 2, then it may happen that neither u nor its harmonic conjugate is Hölder continuous in G with exponent α, as is shown by the example constructed in the proof of Theorem 2 (cf. (1.15)).

2. Proof of Theorem 1.

2.1. Let the assumptions of Theorem 1 be satisfied. Before we make special use of the fact that G is a quasicircle domain, we do some preliminary work along the lines of [**6**, Section 3]. First we recall the following special case of [**6**, Lemma 1].

LEMMA A. *If G is a bounded plane domain, and if u is harmonic in G and continuous in $\bar{G}$, then for every positive t, we have*

$$\sup\{|u(z_1) - u(z_2)| : |z_1 - z_2| \le t, z_1, z_2 \in \bar{G}\}$$
$$= \sup\{|u(z_1) - u(z_2)| : |z_1 - z_2| \le t, z_1 \in \partial G, z_2 \in \bar{G}\}.$$

By Lemma A, to prove any of (1.9)-(1.13) for all $z_1, z_2 \in \bar{G}$, we may assume that $z_1 \in \partial G$ and $z_2 \in G$. When considering a fixed $z_1 \in \partial G$ like this, we shall only make use of the assumption that (1.2) holds for this z_1 and for all $z_2 \in \partial G$. This then also proves the last statement of Theorem 1. From now on, let z_1 be a fixed boundary point of G and set

$$v(z) = u(z) - u(z_1)$$

so that v is harmonic in G, continuous in $\bar{G}$, and $|v(z)| \le \mu(|z - z_1|)$ for $z \in \partial G$. To estimate $|v(z_2)|$ for a fixed but arbitrary $z_2 \in G$, we will find an upper bound for $v(z_2)$. A lower bound is then obtained by applying the same argument to the function $-v$.

By (1.1), we have

$$(2.1) \qquad\qquad \mu(\lambda t) \le 2\lambda\mu(t) \text{ for } t \ge 0, \lambda \ge 1.$$

Suppose that $|z_1 - z_2| \geq d$. Then by (2.1) and (1.2) we have

$$v(z) \leq \mu(|z - z_1|) \leq 2(|z - z_1|/|z_1 - z_2|)\mu(|z_1 - z_2|)$$
$$\leq 2(\operatorname{diam} G)d^{-1}\mu(|z_1 - z_2|)$$

if $z \in \partial G$ and $|z - z_1| \geq |z_1 - z_2|$. If $z \in \partial G$ and $|z - z_1| < |z_1 - z_2|$ then

$$v(z) \leq \mu(|z - z_1|) \leq \mu(|z_1 - z_2|) \leq 2(\operatorname{diam} G)d^{-1}\mu(|z_1 - z_2|).$$

Hence the maximum principle implies that

$$(2.2) \qquad v(z_2) \leq 2(\operatorname{diam} G)d^{-1}\mu(|z_1 - z_2|).$$

The same argument shows that if $\mu(t) = Mt^\alpha$ then (2.2) can be replaced by

$$(2.3) \qquad v(z_2) \leq (\operatorname{diam} G)^\alpha d^{-\alpha}\mu(|z_1 - z_2|).$$

2.2. Suppose then that $|z_1 - z_2| < d$. We denote by $\omega(z_2, \gamma)$ the harmonic measure of the set $\gamma \subset \partial G$ with respect to G at the point z_2. For $r > 0$, we define $E(r) = \{z \in \partial G: |z - z_2| \geq r\}$ and $\Omega(r) = \omega(z_2, E(r))$. We write $\rho = |z_1 - z_2|$.

We define the nondecreasing function $L(r)$ by $L(r) = \mu(\rho)$ for $0 \leq r < \rho$. If $\rho \leq r \leq \operatorname{diam} G$, we set $L(r) = 2r\mu(\rho)/\rho$, except if $\mu(t) = Mt^\alpha$, in which case we set $L(r) = (r/\rho)^\alpha\mu(\rho)$. For $r > \operatorname{diam} G$, we define $L(r) = L(\operatorname{diam} G)$. In each case it follows from (1.2) and (2.1) that

$$(2.4) \qquad v(z) \leq L(|z - z_1|) \text{ for all } z \in \partial G.$$

Let Γ be the component of ∂G containing z_1, let δ be the distance of z_2 from Γ, and let R be the maximum of $|z - z_2|$ for $z \in \Gamma$. Then $\delta \leq \rho$, and we have $R \geq 3d$ by the definition of d. We define the nonincreasing function $\omega(r)$ by $\omega(r) = 10^6$ for $0 \leq r < \delta$, by $\omega(r) = 10^6(\delta/r)^\kappa$ for $\delta \leq r \leq R$ and by $\omega(r) = \omega(R) = 10^6(\delta/R)^\kappa$ for $r > R$. We shall soon show that

$$(2.5) \qquad \Omega(r) \leq \omega(r)$$

for $2\delta < r < R$. Clearly (2.5) then holds for $r \geq R$ also since $\Omega(r)$ is nonincreasing. Further, (2.5) remains valid for $0 \leq r \leq 2\delta$ since $\Omega(r) \leq 1$ for all r and since $\frac{1}{2} < \kappa \leq 1$. Hence (2.5) holds for all $r \geq 0$.

Now we recall some arguments from the proof of [6, Lemma 2]. By the maximum principle, (2.4), (2.5) and integration by parts we obtain, with $R_0 = \operatorname{diam} G$, that

$$
\begin{aligned}
v(z_2) = \int_{\partial G} v(z)\omega(z_2, dz) &\le \int_{\partial G} L(|z - z_1|)\omega(z_2, dz) \\
&\le \int_{\partial G} L(|z - z_2| + \rho)\omega(z_2, dz) = -\int_0^{R_0} L(r + \rho)d\Omega(r) \\
&\le L(\rho)\Omega(0) + \int_0^{R_0} \Omega(r)dL(r + \rho) \\
&\le L(\rho) + \int_0^{R_0 - \rho} \omega(r)dL(r + \rho)
\end{aligned}
$$

(2.6)

since $L(r)$ is constant for $r \ge R_0$. Here $\omega(z_2, dz)$ denotes integration with respect to harmonic measure.

Using the definitions of $L(r)$ and $\omega(r)$ we obtain from (2.6) that

$$
\begin{aligned}
v(z_2) \le L(\rho) + 10^6(L(\delta + \rho) - L(\rho)) + \mu(\rho)\Big\{ 10^6(2/\rho)\int_\delta^R (\delta/r)^\kappa dr \\
+ 10^6(\delta/R)^\kappa(2/\rho)\int_R^{R_0 - \rho} dr \Big\}.
\end{aligned}
$$

(2.7)

If $k < 1$ so that $\kappa < 1$, this gives

$$
(2.8) \qquad v(z_2) \le \mu(\rho)\{2 + 2\cdot10^6 + 4\cdot10^6 R_0 d^{-\kappa}\rho^{\kappa-1}(1 - \kappa)^{-1}\}
$$

since $3d \le R \le R_0$ and $\delta \le \rho$. This together with (2.2) gives (1.9) with, for example, the constant C_1 given after the statement of Theorem 1.

If $k = \kappa = 1$, (2.7) implies that

$$
\begin{aligned}
v(z_2) &\le \mu(\rho)\{2 + 2\cdot10^6[1 + R_0/(3d) + (\delta/\rho)\log(R/\delta)]\} \\
&\le 2\mu(\rho)\{1 + 10^6[1 + \log(R/\rho) + R_0/(3d)]\}
\end{aligned}
$$

(2.9)

which together with (2.2) gives (1.10).

If $\mu(t) = Mt^\alpha$, (2.6) implies that

$$
\begin{aligned}
v(z_2) \le \mu(\rho)\Big\{ 1 + 10^6(2^\alpha - 1) + 10^6\alpha\rho^{-\alpha}\delta^\kappa \int_\delta^R (r + \rho)^{\alpha-1}r^{-\kappa}dr \\
+ 10^6(\delta/R)^\kappa \alpha\rho^{-\alpha} \int_R^{R_0 - \rho} (r + \rho)^{\alpha-1}dr \Big\}.
\end{aligned}
$$

(2.10)

This together with (2.3) gives (1.11)-(1.13) after some calculations with, for example, the constants C_2, A_2 and C_3 mentioned after the statement of Theorem 1. If $R > R_0 - \rho$, the last term in (2.7) and (2.10) should be omitted, but the subsequent upper bounds such as (2.8) and (2.9) are still valid.

2.3. It remains to prove (2.5) for $2\delta < r < R$. With the previous definition of Γ, let D be the component of the complement of Γ in the sphere containing G and hence z_2. For $r > 0$, we denote by $\theta_1(r)$ and $\theta(r)$ the *angular length* of the longest arc contained in $G \cap \{z \colon |z - z_2| = r\}$ and $D \cap \{z \colon |z - z_2| = r\}$, respectively, so that $\theta_1(r) \le \theta(r)$. However, if $\{z \colon |z - z_2| = r\}$ is contained in G or D, we set $\theta_1(r) = \infty$ or $\theta(r) = \infty$, respectively. With the previous definition of $\Omega(r)$ we then have

$$
\begin{aligned}
\text{(2.11)} \qquad \Omega(r) &\le 3\sqrt{2}\exp\{-\pi \int_0^{r/2} (t\theta_1(t))^{-1}\,dt\} \\
&\le 3\sqrt{2}\exp\{-\pi \int_0^{r/2} (t\theta(t))^{-1}\,dt\}
\end{aligned}
$$

by Tsuji's inequality (take $\kappa = 1/2$ in [**7**, Theorem III.67, p. 112] and combine this with [**5**, Lemma 1, p. 123]; cf. also [**6**, Section 5]).

Next we estimate $\theta(r)$. If $\delta < r < R$, where δ and R are as in subsection 2.2, then the intersection $\Gamma \cap \{z \colon |z - z_2| = r\}$ contains at least two points. We can choose points $w_i \in \Gamma$ for $1 \le i \le 4$ such that they follow each other on Γ in the positive or negative order, such that $|w_2 - z_2| = \delta$, $|w_4 - z_2| = R$ and $|w_1 - z_2| = |w_3 - z_2| = r$. Then one of the arcs of $\{z \colon |z - z_2| = r\}$ from w_1 to w_3 contains an arc of $G \cap \{z \colon |z - z_2| = r\}$ of maximal length. We may assume that $w_1 = z_2 + re^{-i\theta}$ and $w_3 = z_2 + re^{i\theta}$ where $0 < \theta < \pi$, and that $\theta(r) \le 2\pi - 2\theta$. If $\theta \ge \pi/2$ then $\theta(r) \le \pi$ and we are happy with this upper bound for $\theta(r)$. So we assume that $\theta < \pi/2$ and look for a lower bound for θ.

Since Γ is a k-quasicircle, (1.7) gives

$$
\frac{2(r - \delta)(R - r)}{2r(\sin\theta)(R + \delta)} \le \frac{|w_1 - w_2||w_3 - w_4| + |w_2 - w_3||w_4 - w_1|}{|w_1 - w_3||w_2 - w_4|} \le \frac{1}{k}.
$$

This implies that

$$
\sin\theta \ge k\left(1 - \frac{\delta}{r}\right)\left(1 - \frac{r + \delta}{R + \delta}\right) > k(1 - x)
$$

where $x = \delta/r + 2r/R$. We have $0 < x \le 2$ for $\delta \le r \le R/2$. Suppose first that $x \le 0.13$. Since $0 \le k(1 - x) \le k \le 1$, this gives

$$
\theta \ge \arcsin(k - kx) = \arcsin k - \arcsin\beta
$$

where $0 \le \beta \le 1$ and

$$
\beta = \sin(\arcsin k)\cos[\arcsin(k - kx)] - \cos(\arcsin k)\sin[\arcsin(k - kx)].
$$

Hence

$$\beta = k(1 - k^2(1 - x)^2)^{1/2} - k(1 - x)(1 - k^2)^{1/2}.$$

It follows that

$$
\begin{aligned}
\beta/k &= (1 - k^2 + k^2(2x - x^2))^{1/2} + (x - 1)(1 - k^2)^{1/2} \\
&\leq (1 - k^2)^{1/2} + k(2x - x^2)^{1/2} + (x - 1)(1 - k^2)^{1/2} \\
&\leq k(2x)^{1/2} + x \leq \sqrt{x}(k\sqrt{2} + \sqrt{x}) \leq (1.775)\sqrt{x}
\end{aligned}
$$

so that $\beta \leq (1.775)k\sqrt{x}$. Hence $\pi\beta/2 \leq (2.79)k\sqrt{x}$ for $0 < k \leq 1$ and $\delta \leq r \leq R/2$ if $x \leq 0.13$. Thus

$$\theta \geq \arcsin k - \pi\beta/2 \geq \arcsin k - (2.79)k\sqrt{x} \geq (\arcsin k)(1 - (2.79)\sqrt{x})$$

since $\arcsin \beta \leq \pi\beta/2$ and $k \leq \arcsin k$. So for $\delta \leq r \leq R/2$ we have

$$
\begin{aligned}
(2.12) \qquad \theta(r) &\leq 2(\pi - \arcsin k)(1 - (2.79)\sqrt{x}) + (5.58)\pi\sqrt{x} \\
&\leq 2(\pi - \arcsin k)(1 + (2.79)\sqrt{x}) \\
&= 2(\pi - \arcsin k)(1 + (2.79)(\delta/r + 2r/R)^{1/2})
\end{aligned}
$$

if $x \leq 0.13$ since $2(\pi - \arcsin k) \geq \pi$. This last upper bound for $\theta(r)$ is valid also if $x > 0.13$ since $\theta(r) \leq 2\pi$.

2.4. By (2.12), we have for $2\delta < r \leq R$,

$$-\pi \int_0^{r/2} \frac{dt}{t\theta(t)} \leq \frac{-\pi}{2(\pi - \arcsin k)} \int_\delta^{r/2} \frac{dt}{t}(1 - (2.79)(\delta/t + 2t/R)^{1/2})$$

since $(1 + y)^{-1} \geq 1 - y$ when $y > 0$. This last integral is

$$
\begin{aligned}
&\geq \log \frac{r}{2\delta} - (2.79) \int_\delta^{r/2} \frac{dt}{t}\{(\delta/t)^{1/2} + (2t/R)^{1/2}\} \\
&= \log \frac{r}{2\delta} - 5.58 - (5.58)\sqrt{r/R} + (5.58)\sqrt{2}(\sqrt{\delta/r} + \sqrt{\delta/R}) \\
&\geq \log(r/\delta) - 11.16 - \log 2
\end{aligned}
$$

since $\sqrt{a + b} \leq \sqrt{a} + \sqrt{b}$ when $a, b \geq 0$. Therefore by (2.11),

$$(2.13) \qquad \Omega(r) \leq 3\sqrt{2}(2e^{11.16})(\delta/r)^\kappa \leq 6 \cdot 10^5 (\delta/r)^\kappa \leq 10^6 (\delta/r)^\kappa$$

for $2\delta < r \leq R$, where κ is as in (1.8). This proves (2.5), and Theorem 1 is proved.

3. Proof of Theorem 2.

Given $k \in (0, 1]$ we note that the domain $G_1 = \{z \colon |\arg z| < \pi - \arcsin k\}$ is bounded by a k-quasicircle. This is proved in detail in [**2**, p. 7-10]. We have $G_1 = \phi(G_2)$ where G_2 is the right half plane and $\phi(z) = z^{1/\kappa}$ for κ given by (1.8). We take the domain G to be $G = \psi(G_1)$ where $\psi(z) = (z - 1)(z + 1)^{-1}$. Then ∂G is a k-quasicircle since ψ is a Möbius transformation, and G, which is the union of two disks and is symmetric about the coordinate axes, is an hourglass-shaped domain with an interior angle π/κ at the boundary points ± 1.

Given $\alpha \in (0, 1]$, we take u to be the harmonic function in G with boundary values $u(z) = |z + 1|^\alpha$ for $z \in \partial G$. Then u satisfies (1.2) for all $z_1, z_2 \in \partial G$ with $\mu(t) = t^\alpha$. Namely,

$$|z_1 + 1|^\alpha \leq (|z_1 - z_2| + |z_2 + 1|)^\alpha \leq |z_1 - z_2|^\alpha + |z_2 + 1|^\alpha$$

since $0 < \alpha \leq 1$, and similarly $|z_2 + 1|^\alpha \leq |z_1 - z_2|^\alpha + |z_1 + 1|^\alpha$.

We take the points w and z_n mentioned in Theorem 2 to be $w = -1 \in \partial G$ and $z_n = -1 + 1/n \in G$ for $n \geq 1$. Then $z_n \to w$ as $n \to \infty$. We have $u(w) = 0$.

We note that $u(z) > 0$ for $z \in G$. To get a lower bound for $u(z_n)$ we observe that the function $v(z) = u(g(z))$, where $g = \psi \circ \phi$, is bounded and harmonic in the right half plane G_2. Now $u(z_n) = v(g^{-1}(z_n))$, and $\varsigma_n = g^{-1}(z_n)$ is real and positive with $\varsigma_n \to 0$ as $n \to \infty$. Further,

$$(3.1) \qquad \varsigma_n = n^{-\kappa}(2 - n^{-1})^{-\kappa} \sim (2n)^{-\kappa}$$

as $n \to \infty$.

The Poisson formula in the right half plane gives

$$v(\varsigma_n) = \frac{\varsigma_n}{\pi} \int_{-\infty}^{\infty} \frac{v(it)}{t^2 + \varsigma_n^2} \, dt$$

where $v(it) = 2^\alpha |t|^{\alpha/\kappa} |1 + (it)^{1/\kappa}|^{-\alpha}$. Hence by (3.1),

$$v(\varsigma_n) \geq An^{-\kappa} \int_0^1 \frac{t^{\alpha/\kappa} dt}{t^2 + n^{-2\kappa}}$$

for some positive absolute constant A and all large n. A straightforward estimate shows that

$$v(\varsigma_n) \geq A_0 n^{-\kappa}(\alpha - \kappa)^{-1} \qquad \text{if } \alpha > \kappa,$$
$$v(\varsigma_n) \geq A_0 n^{-\kappa} \log n \qquad \text{if } \alpha = \kappa,$$

for some positive absolute constant A_0 and all large n. Combining these results and the fact that $\mu(|z_n - w|) = n^{-\alpha}$, we obtain (1.14) and (1.15) for all large n. Theorem 2 is proved.

Mathematical Sciences Research Institute, 1000 Centennial Drive, Berkeley, California 94720
Current Address: Dept. of Mathematics, University of Michigan, Ann Arbor, Michigan 48109-1003

REFERENCES

1. Ahlfors, L.V., *Quasiconformal reflections*, Acta Math. **109** (1963), 291–301.
2. Blevins, D.K., *Some properties of domains bounded by quasiconformal circles*, University of Michigan Ph.D.–thesis, 1972.
3. Blevins, D.K., *Harmonic measure and domains bounded by quasiconformal circles*, Proc. Amer. Math. Soc. **41** (1973), 559–564.
4. Gehring, F.W. and Martio, O., *Quasidisks and the Hardy–Littlewood property*, Complex Variables **2** (1983), 67–78.
5. Hayman, W.K. and Weitsman, A., *On the coefficients and means of functions omitting values*, Math. Proc. Cambridge Phil. Soc. **77** (1975), 119–137.
6. Hinkkanen, A., *Modulus of continuity of harmonic functions*, to appear.
7. Tsuji, M., *Potential theory in modern function theory*, Maruzen, Tokyo, 1959.
8. Zinsmeister, M., *Problèmes de Dirichlet, Neumann, Calderón dans les quasidisques pour les classes Hölderiennes*, to appear.

Note on a theorem of Wolff

BY JOHN L. LEWIS

1. Introduction.

Denote points in $\mathbb{R}^2$ by (x, y) and put $\mathbb{R}^2_+ = \{(x, y) : y > 0\}$. Given p, $1 < p < \infty$, we shall say that u is a weak solution in $\mathbb{R}^2_+$ to

$$(1) \qquad L_p u = \nabla \cdot [|\nabla u|^{(p-2)} \nabla u] = 0,$$

(the so called p Laplacian equation) provided

$$\int_{\mathbb{R}^2_+} |\nabla u|^{(p-2)} \nabla u \cdot \nabla \phi \, dx \, dy = 0,$$

for all $\phi \in C_0^\infty(\mathbb{R}^2_+)$. Here ∇ denotes the gradient and $C_0^\infty(\mathbb{R}^2_+)$ is the space of infinitely differentiable functions on $\mathbb{R}^2_+$ with compact support. Recently Wolff has proved the following theorem.

THEOREM A. *If $p > 2$ then there exist bounded weak solutions of $L_p u = 0$ on $\mathbb{R}^2_+$ such that the set $\{x \in \mathbb{R} : \lim_{y \to 0} u(x, y) \text{ exists}\}$ has measure zero, as well as bounded positive weak solutions of $L_p v = 0$ such that $\{x \in \mathbb{R} : \limsup_{y \to 0} v(x, y) > 0\}$ has measure zero.*

To prove his theorem Wolff brilliantly combines harmonic analysis and P.D.E. techniques to construct certain solutions to (1). The key to his proof is the following existence theorem.

THEOREM B. *If $p > 2$ there exists a bounded Lipschitz function ψ on $\overline{\mathbb{R}}^2_+$ such that $\psi(x + 1, y) = \psi(x, y)$, $\int_{(0,1) \times (0,\infty)} |\nabla \psi|^p dx \, dy < +\infty$, $L_p \psi = 0$ weakly on $\mathbb{R}^2_+$ and*

$$(2) \qquad \int_0^1 \psi(x, 0) dx \neq 0 \quad \text{but } \psi(x, y) \to 0 \text{ as } y \to \infty.$$

To establish Theorem B, he uses a perturbation type argument on a known solution of (1) and then makes some estimates. In this note we prove

THEOREM 1. *Theorems A and B are also true when $1 < p < 2$.*

Wolff (see the discussion following Lemma 1 in [3]) shows that Theorem B implies Theorem A for $1 < p < \infty$. Therefore we shall only prove Theorem B. His proof of Theorem B for $p > 2$ remains valid for $1 < p < 2$ except in one annoying place, where he is unable to make a certain estimate (see the last paragraph of [3]). We also have been unable to make this estimate. However, by a "duality type argument" we are able to reduce the problem of construcing ψ satisfying (2) for $1 < p < 2$ to a similar problem for $p > 2$. We then make liberal use of the Wolff machine to solve this problem.

2. Proof of Theorem 1.

We shall use the same notation as in [3] and shall refer to this paper whenever possible. Therefore the reader is advised to have the paper at hand in order to understand our arguments. Let $1 < p < 2$ and h be a weak solution to $L_p h = 0$ on $\mathbb{R}^2_+$. Then ∇h is a quasiregular mapping from $\mathbb{R}^2_+$ to $\mathbb{R}^2$ so has only isolated zeros. Moreover, it follows from regular ellipticity theory that ∇h is real analytic locally at each point where ∇h is nonzero. Also, ∇h is locally Hölder continuous on $\mathbb{R}^2_+$ with exponent depending only on p. For a proof of these facts see [1, Ch. 11].

Using the above facts, (1), and Green's theorem it is easily seen that there exists a function u on $\mathbb{R}^2_+$ with

$$(3) \qquad u_y = |\nabla h|^{(p-2)} h_x, \qquad u_x = -|\nabla h|^{(p-2)} h_y.$$

Moreover, if $q = p/(p-1) > 2$, then u is a weak solution to

$$\nabla \cdot [|\nabla u|^{(q-2)} \nabla u] = 0 \quad \text{on } \mathbb{R}^2_+.$$

Note that $|\nabla u|^q = |\nabla h|^p$. Since all steps are reversible we claim that in order to prove Theorem 1 it suffices to prove

LEMMA 1. *If $q > 2$, there exists a bounded Lipschitz function u on $\overline{\mathbb{R}}^2_+$ such that $u(x+1, y) = u(x, y)$, $\int_{(0,1)\times(0,\infty)} |\nabla u|^q dx dy < +\infty$, $L_q u = 0$ weakly on $\mathbb{R}^2_+$ and*

$$(4) \qquad \int_0^1 (|\nabla u|^{(q-2)} u_x)(x, a)\, dx \neq 0,$$

for some $a > 0$.

For u as in Lemma 1 we note from Lemma 1.3 in [3] that

$$|u(x, y) - \mu| \leq ce^{-\lambda y} \quad \text{on } \mathbb{R}^2_+,$$

for some $c, \lambda > 0$ and μ real. This inequality and standard estimates (see the proof of Lemma 3.12 in [3] or [2, §3]) imply

$$|\nabla u(x,y)| \leq c_1 e^{-\lambda y} \quad \text{on } \mathbb{R}^2_+$$

for some $c_1 > 0$. Define h relative to u as in (3) with $h(0,1) = 0$. Then h is bounded on $\mathbb{R}^2_+$ since $|\nabla h|^p = |\nabla u|^q$. Also from the divergence theorem, the exponential decay of $|\nabla u|$ as $y \to \infty$, the fact that $L_q u = 0$ on $\mathbb{R}^2_+$, and the periodicity of u in the x variable, we get

$$h(x+1,y) - h(x,y) = \int_x^{x+1} h_t(t,y)dt = \int_x^{x+1} (|\nabla u|^{(q-2)} u_y)(t,y)dt = 0.$$

Hence, $h(x+1,y) = h(x,y)$, (x,y) in $\mathbb{R}^2_+$. Using (3) and (4) we find that $\frac{d}{dy}\int_0^1 h(x,y)dx \neq 0$ at $y = a$. From this discussion we conclude that there exists σ a real number and b near a such that the conclusion of Theorem B is valid for

$$\psi(x,y) = h(x,y+b) - \sigma, \quad (x,y) \text{ in } \mathbb{R}^2_+.$$

Thus in order to prove Theorem B for $1 < p < 2$ it suffices to prove Lemma 1.

To prove Lemma 1 we first use Lemmas 3.1 and 3.3 in [3] to get for a fixed $q > 2$, a unique infinitely differentiable function f on $\mathbb{R}^2$ with the following properties:

(a) $f(0,0) = 1$, $f_x(0,0) = 0$,
(b) $L_q f = 0$ on $\mathbb{R}^2_+$,
(c) $f(x,y) = e^{-y}a(x)$ where $a(x+\gamma) = a(x)$ for some $\gamma > 0$ and

$$(5) \qquad a_{xx} + \Gamma(a, a_x)a = 0 \quad \text{on } \mathbb{R},$$

with

$$\Gamma(a, a_x) = \frac{(2q-3)a_x^2 + (q-1)a^2}{(q-1)a_x^2 + a^2}.$$

Next following Wolff we construct a function v in $C^\infty(\overline{\mathbb{R}}^2_+)$ with $v(x+\gamma,y) = v(x,y)$ and such that $L_q v = 0$ and (4) hold at $a = 0$ with $(0,1)$ replaced by $(0,\gamma)$ and u by $f + \epsilon v$ up to terms in ϵ^2. It will turn out that $f + \epsilon v$ is a close approximation to the function u in Lemma 1. Expanding $L_q(f + \epsilon v)$ and using the fact that $f_x = a_x$, $f_y = -a$, we get

$$(6) \qquad TV = \nabla \cdot (A\nabla v) = 0,$$

where A is the symmetric matrix

$$(a_x^2 + a^2)^{(q-4)/2} e^{-(q-2)y} \begin{pmatrix} a^2 + (q-1)a_x^2 & -(q-2)a\,a_x \\ -(q-2)a\,a_x & (q-1)a^2 + a_x^2 \end{pmatrix}$$

and ∇v is written in column form. Observe from the uniqueness of f that $f(x,y) = f(-x,y)$. Thus for a fixed $y \geq 0$,

$$(7) \qquad 2\int_0^\gamma (|\nabla f|^{(q-2)} f_x)(x,y)\,dx = \int_{-\gamma}^\gamma (|\nabla f|^{(q-2)} f_x)(x,y)\,dx = 0.$$

Expanding (4) with u replaced by $f + \epsilon v$ and using (7) we find that

$$(8) \qquad \int_0^\gamma (v_x v_y) A(x,0) \begin{pmatrix} 1 \\ 0 \end{pmatrix} dx \neq 0,$$

is necessary in order that (4) holds for $f + \epsilon v$ at $a = 0$ up to terms in ϵ^2. In (8), $(v_x v_y)$ denotes a 1×2 row matrix.

To construct v let

$$\nu^*(x) = A(x,0) \begin{pmatrix} 0 \\ -1 \end{pmatrix}, \qquad \xi(x) = A(x,0) \begin{pmatrix} 1 \\ 0 \end{pmatrix},$$

and define real valued functions μ, τ, on $\mathbb{R}$ by

$$(9) \qquad b\nu^* + \xi = \mu \left[\nu^* + \tau \begin{pmatrix} 1 \\ 0 \end{pmatrix} \right],$$

where b is a positive constant to be chosen later. Thus,

$$(10) \qquad \mu = \frac{(q-2)a\,a_x}{a_x^2 + (q-1)a^2} + b,$$

$$\mu\tau = (a_x^2 + a^2)^{(q-4)/2}[(q-1)a_x^2 + a^2 + (b-\mu)(q-2)a\,a_x].$$

Observe that if b is large enough, then $\mu > 0$ on $[0,\gamma]$ and hence $\tau \in C^\infty(\mathbb{R})$. For b satisfying the above requirements we shall need the following lemma which is Lemma 3.7 in [3].

LEMMA A. *Let E be the set of all real valued functions g on $\mathbb{R}$ for which there exists G in $C^\infty(\overline{\mathbb{R}}_+^2)$ satisfying*

(i) $G = g$ *and* $\frac{\partial G}{\partial \nu^*} - \frac{\partial}{\partial x}(\tau G) = 0$ *on* $\mathbb{R}$,
(ii) $TG = 0$ *and* $G(x+\gamma, y) = G(x,y)$ *on* $\mathbb{R}_+^2$,
(iii) *for a fixed* $\beta > q - 2$,

$$\int_{(0,\gamma)\times(0,\infty)} (|\nabla G|^2 e^{-(q-2)y} + |G|^2 e^{-\beta y})\,dx\,dy < +\infty.$$

Then E is finite dimensional and if $\phi \in C^\infty(\mathbb{R})$ is such that $\phi(x+\gamma) = \phi(x)$ and $\int_0^\gamma \phi g\,dx = 0$ for all $g \in E$, there is a Φ satisfying (ii), (iii), with G replaced by Φ. Moreover,

$$\frac{\partial \Phi}{\partial \nu^*} + \tau \frac{\partial \Phi}{\partial x} = \phi \quad \text{on } \mathbb{R}.$$

Let Φ be as in Lemma A. Then from (iii) it follows that

$$\int_0^\gamma (\Phi_x \Phi_y) A(x,R) \binom{0}{1}\,dx \to 0$$

as $R \to \infty$ through a certain sequence. Using this fact, (ii), and the divergence theorem we get $\int_0^\gamma \frac{\partial \Phi}{\partial \nu^*}\,dx = 0$. Next we note that if there exists $\phi \in C^\infty(\mathbb{R})$ with period γ, $\int_0^\gamma \phi g\,dx = 0$ for all $g \in E$, and $\int_0^\gamma \mu\phi\,dx \neq 0$, then by Lemma A with $v = \Phi$,

$$0 \neq \int_0^\gamma \mu\phi\,dx = \int_0^\gamma \mu \left(\frac{\partial v}{\partial \nu^*} + \tau \frac{\partial v}{\partial x} \right)\,dx$$

$$= \int_0^\gamma \left(\frac{\partial v}{\partial \xi} + b\frac{\partial v}{\partial \nu^*} \right)\,dx = \int_0^\gamma \frac{\partial v}{\partial \xi}\,dx.$$

Thus in this case v satisfies (8).

Writing μ as $\mu_1 + \mu_2$ where $\mu_1 \in E$ and μ_2 is orthogonal to E with respect to the usual inner product on $(0,\gamma)$, we see that the only way such ϕ cannot exist is if $\mu \in E$. That is there exists G as in Lemma A, with

$$(11) \qquad \frac{\partial G}{\partial \nu^*} = \frac{d}{dx}(\mu\tau), \quad \text{and } G = \mu \text{ on } \mathbb{R}.$$

We show that (11) leads to a contradiction.

Let x_0 be a point where μ assumes an absolute minimum on $\mathbb{R}$. Then from (10) we see at x_0 that

$$0 = \frac{d\mu}{dx} = -\left[a_x^2 + (q-1)a^2\right]^{-2}(q-2)(a\,a_{xx} - a_x^2)(a_x^2 - (q-1)a^2).$$

Now from (5) it follows that $aa_{xx} \neq a_x^2$ on $\mathbb{R}$ so $\frac{d\mu}{dx} = 0$ only if $a_x^2 = (q-1)a^2$ at x_0 (from ordinary differential equation theory a and a_x cannot vanish simultaneously on $\mathbb{R}$). Thus at x_0, $a_x = -\sqrt{q-1}\,a$, $\mu - b = -(q-2)/2\sqrt{q-1}$, and from (5)

$$a_{xx} = \frac{-2(q-1)^2}{(q-1)^2 + 1}a.$$

If k_1, k_2, are two real valued functions on $\mathbb{R}$, we write $k_1 \sim k_2$ to mean that k_1/k_2 is positive at x_0.

From (10) and the above we find at x_0 that

$$
\begin{aligned}
\tfrac{d}{dx}(\mu\tau) &\sim \{(q-4)[a_x a_{xx} + a\,a_x][(q-1)a_x^2 + a^2 + (b-\mu)(q-2)a\,a_x] \\
&\quad + (a_x^2 + a^2)[2(q-1)a_x a_{xx} + 2a\,a_x + (b-\mu)(q-2)(a_x^2 + a\,a_{xx})]\} \\
&\sim \left\{ \tfrac{(q-4)(q-2)q\sqrt{q-1}}{(q-1)^2+1}[(q-1)^2 + 1 - \tfrac{1}{2}(q-2)^2] \right. \\
&\quad \left. + \tfrac{q}{\sqrt{q-1}}\left[\tfrac{4(q-1)^4}{(q-1)^2+1} - 2(q-1) + \tfrac{(q-2)^2}{2}\tfrac{((q-1)^3+(q-1)-2(q-1)^2)}{(q-1)^2+1}\right]\right\} \\
&\sim \tfrac{(q-2)q\sqrt{q-1}}{(q-1)^2+1}\{(q-4)[(q-1)^2 + 1 - \tfrac{1}{2}(q-2)^2] \\
&\quad + 4(q-1)^2 + 2(q-1) + 2 + \tfrac{1}{2}(q-2)^3\}.
\end{aligned}
$$

The last term is clearly positive if $q > 4$ and simple estimates show that it continues to remain positive if $2 < q < 4$. Thus, at x_0, $\frac{d}{dx}(\mu\tau) > 0$. To obtain a contradiction we note from Lemma 3.8 in [3] that x_0 is an absolute minimum of G on $\overline{\mathbb{R}}_+^2$. Using this fact and (11) it follows that at x_0

$$0 \le \frac{\partial G}{\partial y} \sim -\frac{\partial G}{\partial \nu^*} = -\frac{d}{dx}(\mu\tau) < 0.$$

From this contradiction we conclude by way of Lemma A that there exists $v \in C^\infty(\overline{\mathbb{R}}_+^2)$ satisfying (6) and (8) with $v(x + \gamma, y) = v(x, y)$, $y > 0$.

Finally we show that if k is a weak solution to $L_q k = 0$ on $\mathbb{R}_+^2$ with $k = f + \epsilon v$ on $\mathbb{R}$, and

$$\int_{(0,\gamma)\times(0,\infty)} |\nabla k|^q dx dy < +\infty,$$

then Lemma 1 holds for $u(x, y) = k(\gamma x, \gamma y)$ and $\epsilon > 0$ sufficiently small. The existence of k follows from the periodicity of f, v in the x variable and standard arguments in the calculus of variations (see [3], §1). It remains to prove (4). To do so let c denote a positive absolute constant, not necessarily the same at each occurence, and put $Q = (0, \gamma) \times (0, \delta)$, where δ, $0 < \delta < 1$ is to be chosen later. Then

$$\left| \int_Q [|\nabla k|^{(q-2)} k_x - |\nabla f + \epsilon \nabla v|^{(q-2)} (f + \epsilon v)_x] dx dy \right|^2$$

$$(12) \qquad \leq c \left[\int_Q (|\nabla k|^{(q-2)} + |\nabla f + \epsilon \nabla v|^{(q-2)}) |\nabla (f + \epsilon v) - \nabla k| dx dy \right]^2$$

$$\leq c \int_Q (|\nabla k|^{(q-2)} + |\nabla f + \epsilon \nabla v|^{(q-2)}) |\nabla (f + \epsilon v) - \nabla k|^2 dx dy,$$

where we have used Schwarz's inequality and the fact tht $q > 2$ to get the last integral. To estimate this integral we note that Wolff's arguments (with g replaced by k) from Lemma 3.16 on all apply to our v. He shows (see the inequalities following (3.20)) that

$$(13)$$
$$\int_{(0,\gamma) \times (0,\infty)} [|\nabla (f + \epsilon v - k)|^q + |\nabla (f + \epsilon v - k)|^2 |\nabla (f + \epsilon v)|^{(q-2)}] dx dy \leq c \epsilon^\tau,$$

for some $\tau = \tau(q)$, $2 < \tau \leq 4$. Now either

$$|\nabla k|(x) \leq 2 \max_{x \in Q} |\nabla (f + \epsilon v)| \leq c \min_{x \in Q} |\nabla (f + \epsilon v)|,$$

or

$$|\nabla k|(x) \leq 2 |\nabla (f + \epsilon v - k)|(x),$$

when $x \in Q$. In either case we see from (12) and (13) that

$$\left| \int_Q [|\nabla k|^{(q-2)} k_x - |\nabla f + \epsilon \nabla v|^{(q-2)} (f + \epsilon v)_x] dx dy \right|^2 \leq c \epsilon^\tau.$$

Using (7) and the above inequality it follows that

$$\left| \int_Q \left[|\nabla k|^{(q-2)} k_x - \epsilon (v_x v_y) A(x, y) \binom{1}{0} \right] dx dy \right|$$

$$\leq \left| \int_Q [|\nabla k|^{(q-2)} k_x - |\nabla f + \epsilon \nabla v|^{(q-2)} (f + \epsilon v)_x] dx dy \right|$$

$$+ \left| \int_Q \left[|\nabla f + \epsilon \nabla v|^{(q-2)} (f + \epsilon v)_x - \epsilon (v_x v_y) A(x, y) \binom{1}{0} \right] dx dy \right| \leq c \epsilon^{\tau/2} + c \epsilon^2.$$

From this inequality and (8) we deduce that if ϵ, δ are sufficiently small, then for some a, $0 < a < \delta/\gamma$,

$$0 \neq \int_0^\gamma |\nabla k|^{(q-2)} k_x(x, a\gamma) dx = \gamma^{2-q} \int_0^1 |\nabla u|^{(q-2)} u_x(x, a) dx.$$

Hence (4) holds and the proof of Lemma 1 is complete. Theorems A and B follow from Lemma 1 in the way mentioned earlier.

Department of Mathematics, University of Kentucky at Lexington, Lexington, KY 40506

REFERENCES

1. Gilbarg, D. and Trudinger, N., "Elliptic Partial Differential Equations of Second Order," Springer Verlag, 1977.
2. Lewis, J., *Regularity of the derivatives of solutions to certain degenerate elliptic equations*, Indiana University Mathematical Journal, **32**(6) (1983), 849–858.
3. Wolff, T., *Gap series constructions for the p Laplacian*, to appear.

Bloch and normal functions on general planar regions

BY DAVID MINDA

Abstract. Let Ω be a hyperbolic region in the complex plane and λ_Ω the density of the hyperbolic metric on Ω. Set $\delta_\Omega(z) = \text{dist}(z, \partial\Omega)$; $1/\delta_\Omega$ is called the quasihyperbolic density on Ω. Roughly speaking, we show that holomorphic functions cannot distinguish between λ_Ω and $1/\delta_\Omega$ while meromorphic functions sometimes can. More precisely, for a holomorphic function f on Ω the quantities $|f'(z)|/\lambda_\Omega(z)$ and $|f'(z)|\delta_\Omega(z)$ are both either uniformly bounded on Ω (that is, f is a Bloch function) or unbounded. With the Euclidean derivative $|f'|$ replaced by the spherical derivative $f^\sharp = |f'|/(1+|f|^2)$, Lehto and Virtanen have observed that the analogous result is generally false. However, we characterize those regions for which there exists a finite constant $n = n(\Omega)$ such that $f^\sharp(z)\delta_\Omega(z) \leq f^\sharp(z)/\lambda_\Omega(z) \leq nf^\sharp(z)\delta_\Omega(z)$, $z \in \Omega$, for any meromorphic function on Ω. In addition, we present another characterization of Bloch functions. A holomorphic function f on Ω is not a Bloch function if and only if there is a sequence $\{z_n\}_{n=1}^\infty$ in Ω and a sequence $\{\rho_n\}_{n=1}^\infty$ of positive numbers such that $\rho_n/\delta_\Omega(z_n) \to 0$ and $f(z_n + \rho_n\varsigma) - f(z_n) \to a\varsigma$, where $|a| = 1$, locally uniformly on C.

1. **Introduction.** A holomorphic function f on the unit disk D is called a Bloch function if

$$(1) \qquad \sup\{(1 - |z|^2)|f'(z)| : z \in D\} < \infty.$$

A meromorphic function f on D is called a normal function if

$$(2) \qquad \sup\{(1 - |z|^2)f^\sharp(z) : z \in D\} < \infty,$$

where $f^\sharp = |f'|/(1 + |f|^2)$ is the spherical derivative of f. We discuss two natural ways to extend both of these notions to more general regions in the complex plane C.

First, note that $\lambda_D(z) = 1/(1 - |z|^2)$ is the density of the hyperbolic metric on D. Suppose Ω is any hyperbolic region in C; that is, $C\backslash\Omega$ contains at least two points. The density λ_Ω of the hyperbolic metric on Ω is

1980 Mathematics Subject Classification, Primary 30D45, 30C99

Key words and phrases. Hyperbolic metric, quasihyperbolic metric

This work was supported in part by NSF Grant DMS-8521158.

determined from $\lambda_\Omega(h(z))|h'(z)| = 1/(1 - |z|^2)$, where $h : D \to \Omega$ is any holomorphic universal covering projection of D onto Ω. Let $B(\Omega)$ denote the set of all holomorphic functions f on Ω for which

$$(3) \qquad \|f\|_B = \sup\left\{\frac{|f'(z)|}{\lambda_\Omega(z)} : z \in \Omega\right\} < \infty.$$

Similarly, let $N(\Omega)$ be the collection of all meromorphic functions f on Ω such that

$$(4) \qquad \|f\|_N = \sup\left\{\frac{f^\sharp(z)}{\lambda_\Omega(z)} : z \in \Omega\right\} < \infty.$$

For $\Omega = D$ these definitions of Bloch and normal functions obviously coincide with (1) and (2), respectively. In fact, (3) and (4) are the customary definitions of Bloch and normal functions on hyperbolic regions; these definitions actually make sense on any hyperbolic Riemann surface.

There is another natural way to extend the notions of Bloch and normal function to any proper subregion of C. Let $\delta_\Omega(z) = \text{dist}(z, \partial\Omega)$; this is the radius of the largest disk in Ω with center z. Note that $\delta_D(z) = 1 - |z|$. The quantity $1/\delta_\Omega$ is called the density of the quasihyperbolic metric on Ω. It has many interesting applications in geometric function theory; for example, see [5] and [6]. A holomorphic function f on Ω is called quasi-Bloch, written $f \in QB(\Omega)$, provided

$$(5) \qquad \|f\|_{QB} = \sup\{\delta_\Omega(z)|f'(z)| : z \in \Omega\} < \infty.$$

Analogously, a meromorphic function f on Ω is called quasinormal, written $f \in QN(\Omega)$, if

$$(6) \qquad \|f\|_{QN} = \sup\{\delta_\Omega(z)f^\sharp(z) : z \in \Omega\} < \infty.$$

For $\Omega = D$ it is clear that (5) and (6) are equivalent to (1) and (2), respectively.

For an arbitrary hyperbolic region Ω there is a simple relationship between the hyperbolic and quasihyperbolic metric, namely, $\lambda_\Omega \leq 1/\delta_\Omega$ [7, p. 45]. This inequality yields $\|f\|_{QB} \leq \|f\|_B$ so that $B(\Omega) \subset QB(\Omega)$. Similarly, $\|f\|_{QN} \leq \|f\|_N$ and $N(\Omega) \subset QN(\Omega)$. On the other hand, for any hyperbolic simply connected region Ω, we have $1/4\delta_\Omega \leq \lambda_\Omega$ [7, p. 45]. Thus, $\|f\|_B \leq 4\|f\|_{QB}$ so the Bloch and quasi-Bloch norms are equivalent on a hyperbolic simply connected region Ω; in particular, $B(\Omega) = QB(\Omega)$. Similarly, the normal and quasi-normal norms are equivalent and $N(\Omega) = QN(\Omega)$ for a simply connected region Ω. For an arbitrary hyperbolic region Ω there does not exist a positive constant $c = c(\Omega)$ such that $c/\delta_\Omega \le \lambda_\Omega$. For instance, if $D^* = D\backslash\{0\}$, then

$$\lambda_{D^*}(z) = \frac{1}{2|z|\log(1/|z|)}$$

and $\delta_{D^*}(z) = |z|$ for $0 < |z| \le 1/2$ so that $\lambda_{D^*}(z)\delta_{D^*}(z) \to 0$ as $z \to 0$. Beardon and Pommerenke [2] and Pommerenke [11] have obtained a necessary and sufficient condition for the existence of such a positive constant c. Thus, at first glance, it is not obvious whether the equalities $B(\Omega) = QB(\Omega)$ and $N(\Omega) = QN(\Omega)$, or the equivalence of the norms, extend from simply connected regions to arbitrary hyperbolic regions.

Lehto and Virtanen [8, Thm. 9] showed that a meromorphic function f on D^* belongs to $N(D^*)$ if and only if f has a pole or a removable singularity at the origin. On the other hand, it is easy to exhibit a function in $QN(D^*)$ with essential singularity at the origin (see Section 5 below), so $N(\Omega)$ can be a proper subset of $QN(\Omega)$. However, we can characterize when the two norms $\|f\|_N$ and $\|f\|_{QN}$ are equivalent. There is a finite constant $n = n(\Omega)$ such that $\|f\|_N \le n\|f\|_{QN}$ for all meromorphic functions f on Ω if and only if Ω is uniformly perfect, that is, there is a uniform bound on the modulus of any annulus in Ω which separates the boundary of Ω. This is equivalent to the existence of $c = c(\Omega) > 0$ such that $c/\delta_\Omega \le \lambda_\Omega$ [2].

In view of the foregoing, our result that $B(\Omega) = QB(\Omega)$ in general is not without interest. More precisely we show that there is an absolute constant $b \le 16/\sqrt{3}$ such that $\|f\|_B \le b\|f\|_{QB}$ for any holomorphic function f on any hyperbolic region Ω. Crudely put, holomorphic functions cannot distinguish between the hyperbolic and quasihyperbolic metric, while meromorphic functions sometimes do. Another main result, a characterization of $B(\Omega)$, relies upon the identity $B(\Omega) = QB(\Omega)$. We show that $f \notin B(\Omega)$ if and only if there exists a sequence $\{z_n\}_{n=1}^\infty$ in Ω and a sequence $\{\rho_n\}_{n=1}^\infty$ of positive numbers such that $\rho_n/\delta_\Omega(z_n) \to 0$ and $f(z_n+\rho_n\varsigma)-f(z_n) \to a\varsigma$, where $|a| = 1$, locally uniformly on C. This is analogous to a characterization of normal functions on D by Lohwater and Pommerenke [9]. Their work readily extends to $QN(\Omega)$ but not to $N(\Omega)$. Very roughly speaking, our characterization is that $f \notin B(\Omega)$ if and only if the Riemann image surface of f asymptotically contains the image surface of a Euclidean motion.

This makes more precise the fact that a holomorphic function is not Bloch if and only if its Riemann image surface contains arbitrarily large unramified disks. The result of Lohwater and Pommerenke is that $f \notin N(D)$ if and only if the Riemann image surface of f asymptotically contains the image surface of a nonconstant Yosida function [3, Thm. 6]. Recall that a Yosida function is a meromorphic function on C with uniformly bounded spherical derivative.

I would like to thank the referee for carefully reading the paper and making a number of useful suggestions, especially for conjecturing Theorem 4 and outlining its proof.

2. Bloch and quasi-Bloch functions. Let f be holomorphic in Ω. For $z \in \Omega$ let $r(z, f)$ be the radius of the largest unramified disk about $f(z)$ in the Riemann image surface of f. Note that $r(z, f) = 0$ if and only if $f'(z) = 0$. Set $r(f) = \sup\{r(z, f) : z \in \Omega\}$. For the unit disk it is well known that $\|f\|_B < \infty$ if and only if $r(f) < \infty$. This result readily extends to arbitrary hyperbolic regions by applying the result for the disk D to the function $f \circ h$, where $h : D \to \Omega$ is a holomorphic universal covering projection.

LEMMA 1. *Suppose $\Omega \neq C$ is a region and f is holomorphic in Ω. Then $r(z, f) \leq 4\delta_\Omega |f'(z)|$ for $z \in \Omega$.*

PROOF: Fix $a \in \Omega$ and set $r = r(a, f)$, $b = f(a)$. There is nothing to prove if $r = 0$, so we may assume $r > 0$. Then there is a branch $g = f^{-1}$ of the inverse function of f which is defined in the disk $D(b, r)$, satisfies $g(b) = a$ and maps $D(b, r)$ into Ω. Therefore,

$$h(w) = \frac{g(b + rw) - g(b)}{rg'(b)}$$

is a normalized univalent function in D. The Koebe $1/4$-theorem implies that $h(D) \supset D(0, 1/4)$, so that

$$\Omega \supset g(D(b, r)) \supset D(a, r|g'(b)|/4) = D(a, r/4|f'(a)|).$$

Hence,

$$\frac{r}{4|f'(a)|} \leq \delta_\Omega(a),$$

which is the desired result.

THEOREM 1. *For any hyperbolic region* Ω, $QB(\Omega) = B(\Omega)$. *More precisely, for any holomorphic function* f *on* Ω,

$$\|f\|_{QB} \leq \|f\|_B \leq \frac{16}{\sqrt{3}}\|f\|_{QB}.$$

PROOF: The left-hand inequality is a consequence of the inequality $\delta_\Omega \leq 1/\lambda_\Omega$ [7, p. 45]. Now, suppose $\|f\|_{QB} < \infty$. Lemma 1 implies that $r(f) \leq 4\|f\|_{QB}$. With $\tau = 4\|f\|_{QB}$ [10, Thm. 5] gives

$$\frac{|f'(z)|}{\lambda_\Omega(z)} \leq \frac{4\tau}{\sqrt{3}}.$$

This establishes the right-hand inequality.

Timoney [12] proved that $f \in B(D)$ if and only if there exists a set $E \subset C$ such that $C\backslash E$ does not contain arbitrarily large disks and $\sup\{(1-|z|^2)|f'(z)| : z \in f^{-1}(E)\} < \infty$. This result readily extends to an arbitrary hyperbolic region Ω by making use of the hyperbolic metric. It is reasonable to inquire whether a similar result holds with the quasi-hyperbolic metric in place of the hyperbolic metric.

THEOREM 2. *Let* Ω *be a hyperbolic region in* C. *Suppose* E *is a subset of* C *and there exists a finite* $R > 0$ *such that any disk in* $C\backslash E$ *has radius at most* R. *If* f *is holomorphic on* Ω *and* $M = \sup\{\delta_\Omega(z)|f'(z)| : z \in f^{-1}(E)\} < \infty$, *then* $f \in B(\Omega)$.

PROOF: It suffices to show that $r(f) < \infty$. Lemma 1 implies $r(z, f) \leq 4M$ for $z \in f^{-1}(E)$. We shall show that $r(z, f) \leq 4M + R$ for all $z \in \Omega$. This inequality is trivial if $z \in f^{-1}(E)$. Suppose $z \in \Omega\backslash f^{-1}(E)$ and $r = r(z, f)$. We show that $r > 4M + R$ is impossible. There is a simply connected region $\Delta \subset \Omega$ such that $f|\Delta$ is injective and $f(\Delta) = \{w : |w - f(z)| < r\}$. Because $r > R$, there exists $a \in \Delta$ with $f(a) \in E$ and $|f(z) - f(a)| \leq R$. Since $r > 4M + R$, there exists $\varepsilon > 0$ such that $D(f(a), 4M + \varepsilon) \subset D(f(z), r) = f(\Delta)$. This yields $r(a, f) \geq 4M + \varepsilon$, a contradiction. Thus $r(f) \leq 4M + R$ and $f \in B(\Omega)$.

3.   The basic lemma. Lohwater and Pommerenke [9] established a necessary and sufficient condition for a function to be non-normal. The following result is basically an adaptation of their idea, but applied to the Euclidean derivative rather than the spherical derivative.

LEMMA 2. *Suppose $\{\Omega_m\}_{m=1}^{\infty}$ is a sequence of regions in the complex plane such that $\overline{\Omega}_m$ is compact for all m. Assume f_m is holomorphic on $\overline{\Omega}_m$ and $w_m \in \Omega_m$ satisfies*

$$\delta_{\Omega_m}(w_m)|f_m'(w_m)| = K_m = \max_{z \in \Omega_m} \delta_{\Omega_m}(z)|f_m'(z)|.$$

Set $r_m = \delta_{\Omega_m}(w_m)/K_m$ and $g_m(\varsigma) = f_m(w_m + r_m\varsigma) - f_m(w_m)$. If $K_m \to \infty$, then $r_m/\delta_{\Omega_m}(w_m) \to 0$ and there is a subsequence $\{g_{m_n}\}_{n=1}^{\infty}$ which converges locally uniformly on C to $g(\varsigma) = a\varsigma$, where $|a| = 1$.

PROOF: Set $\delta_m = \delta_{\Omega_m}$. Since $r_m/\delta_m(w_m) = 1/K_m$ and $K_m \to \infty$, it is clear that $r_m/\delta_m(w_m) \to 0$. Observe that g_m is defined for $|\varsigma| < K_m$ since $\varsigma \mapsto w_m + r_m\varsigma$ maps the disk $D(0, K_m)$ onto $D(w_m, \delta_m(w_m)) \subset \Omega_m$ and that for all m

$$g_m(0) = 0, \quad |g_m'(0)| = r_m|f_m'(w_m)| = 1.$$

Now, we show that the sequence $\{g_m'\}_{m=1}^{\infty}$ is locally uniformly bounded. Fix any compact subset E of C. Because $K_m \to \infty$, there exists $M = M(E)$ such that $E \subset D(0, K_m)$ for all $m \geq M$. For $\varsigma \in E$ and $m \geq M$ we have

$$
\begin{aligned}
(7) \quad |g_m'(\varsigma)| &= r_m|f_m'(w_m + r_m\varsigma)| \leq \tfrac{r_m K_m}{\delta_m(w_m + r_m\varsigma)} \\
&= \tfrac{\delta_m(w_m + r_m\varsigma)^{-1}}{\delta_m(w_m)} \leq \left(1 - |1 - \tfrac{\delta_m(w_m + r_m\varsigma)}{\delta_m(w_m)}|\right)^{-1} \\
&\leq \left(1 - \tfrac{r_m|\varsigma|}{\delta_m(w_m)}\right)^{-1} = \left(1 - \tfrac{|\varsigma|}{K_m}\right)^{-1}.
\end{aligned}
$$

Here we have used the fact that $w_m, w_m + r_m\varsigma \in \Omega_m$ imply

$$|\delta_m(w_m) - \delta_m(w_m + r_m\varsigma)| \leq |w_m - (w_m + r_m\varsigma)| = r_m|\varsigma|, \quad |\varsigma| < K_m.$$

Since $K_m \to \infty$ and $|\varsigma|$ is uniformly bounded on E, the inequality (7) shows that $\{g_m'\}_{m=1}^{\infty}$ is uniformly bounded on E. Because $g_m(0) = 0$ for all m, it easily follows that $\{g_m\}_{m=1}^{\infty}$ is also locally uniformly bounded on C. By Montel's Theorem [4, p. 153], $\{g_m\}_{m=1}^{\infty}$ is a normal family. Thus, there exists a subsequence $\{g_{m_n}\}_{n=1}^{\infty}$ which converges locally uniformly on C to an entire function g. Clearly, $g(0) = 0$ and $|g'(0)| = 1$. The inequality (7) yields $|g'(\varsigma)| \leq 1$ for all $\varsigma \in C$. Liouville's Theorem implies that g' is constant. Hence, $g'(\varsigma) = a$, where $|a| = 1$, and so $g(\varsigma) = a\varsigma$.

4. A characterization of Bloch functions. We establish a new characterization of Bloch functions which relies on the equality $QB(\Omega) = B(\Omega)$.

THEOREM 3. *Let Ω be a hyperbolic region in C and let f be holomorphic on Ω. Then $f \notin B(\Omega)$ if and only if there exists a sequence $\{z_n\}_{n=1}^{\infty}$ in Ω and a sequence $\{\rho_n\}_{n=1}^{\infty}$ of positive numbers with $\rho_n/\delta_\Omega(z_n) \to 0$ and $f(z_n + \rho_n\varsigma) - f(z_n) \to a\varsigma$, where $|a| = 1$, locally uniformly on C.*

PROOF: First, we establish the sufficiency. Thus, we are assuming the existence of sequences $\{z_n\}_{n=1}^{\infty}$ and $\{\rho_n\}_{n=1}^{\infty}$ as in the statement of the theorem. Set $g_n(\varsigma) = f(z_n + \rho_n\varsigma) - f(z_n)$ and $g(\varsigma) = a\varsigma$. Then

$$1 = |g'(0)| = \lim_{n\to\infty} |g_n'(0)| = \lim_{n\to\infty} \frac{\rho_n}{\delta_\Omega(z_n)} \cdot \delta_\Omega(z_n)|f'(z_n)|.$$

Since $\rho_n/\delta_\Omega(z_n) \to 0$, we must have $\delta_\Omega(z_n)|f'(z_n)| \to \infty$. Therefore, $\|f\|_{QB} = \infty$, or $f \notin B(\Omega)$.

Next, we demonstrate the necessity. We are supposing that $\|f\|_{QB} = \infty$. Let $\{\Omega_m\}_{m=1}^{\infty}$ be a sequence of regions such that $\overline{\Omega}_m$ is a compact subset of Ω for all m, $\overline{\Omega}_m \subset \Omega_{m+1}$ for all m and $\bigcup_{m=1}^{\infty} \Omega_m = \Omega$. Set $K_m = \max\{\delta_{\Omega_m}(z)|f'(z)| : z \in \Omega_m\}$. Because f is holomorphic on $\overline{\Omega}_m$ and $\delta_{\Omega_m}(z) = 0$ for all $z \in \partial\Omega_m$, there is a point $w_m \in \Omega_m$ with $\delta_{\Omega_m}(w_m)|f'(w_m)| = K_m$. Since $\Omega_m \subset \Omega_{m+1}$ implies $\delta_{\Omega_m}(z) \le \delta_{\Omega_{m+1}}(z)$ for all $z \in \Omega_m$, it follows that $K_m \le K_{m+1}$ for all m. We claim that K_m increases to ∞. If not, then there exists K such that $\delta_{\Omega_m}(z)|f'(z)| \le K$ for all $z \in \Omega_m$ and all m. Fix $c \in \Omega$. Then $c \in \Omega_m$ for all m sufficiently large and $\delta_{\Omega_m}(c) \to \delta_\Omega(c)$. Hence, we conclude that $\delta_\Omega(c)|f'(c)| \le K$ for all $c \in \Omega$, which contradicts $f \notin QB(\Omega)$. Therefore, $K_m \to \infty$. Then Lemma 2 with $f_m = f$ for all m yields a sequence $\{z_n = w_{m_n}\}_{n=1}^{\infty}$ in Ω and a sequence $\{\rho_n = r_{m_n}\}_{n=1}^{\infty}$ of positive numbers such that $\rho_n/\delta_{\Omega_{m_n}}(z_n) \to 0$ and $f(z_n + \rho_n\varsigma) - f(z_n) \to a\varsigma$, $|a| = 1$, locally uniformly on C. From $\delta_{\Omega_{m_n}}(z_n) \le \delta_\Omega(z_n)$ we obtain $\rho_n/\delta_\Omega(z_n) \to 0$. This completes the proof.

5. Normal and quasi-normal functions. We begin by constructing an example of a function which is quasi-normal, but not normal, on the punctured disk D^*. Later, this example will be used in the proof that characterizes regions for which the norms $\|f\|_N$ and $\|f\|_{QN}$ are equivalent. Let g be a nonconstant, doubly-periodic, meromorphic function on C with periods 1 and $2\pi i$. Then $g^\sharp$ is doubly-periodic, continuous and real-valued on C, so it is bounded, say $g^\sharp(w) \le K$ for $w \in C$. Now, $f(z) = g(\log z)$ defines a single-valued meromorphic function on $C\backslash\{0\}$ with an essential singularity at the origin. This insures that $f \notin N(D^*)$ [8]. On the other hand, $f^\sharp(z) = g^\sharp(\log z)/|z|$ gives $|z|f^\sharp(z) \le K$ for $z \ne 0$, so $f \in QN(D^*)$.

Next, we investigate this function in more detail. Since $g^\sharp$ vanishes only at multiple zeros or poles of g, there exists V_0 such that $g^\sharp(w)$ does not vanish on $\{w : \operatorname{Im}(w) = V_0\}$. There is no harm in assuming that $V_0 = 0$. Because $g^\sharp$ has period 1, there is a (possibly different) constant K such that $g^\sharp(u) \geq 1/K$ for all $u \in R$. Then for $x > 0$,

$$x f^\sharp(x) = g^\sharp(\ln(x) + 2\pi n i) = g^\sharp(\ln(x)) \geq 1/K.$$

This function f, which is meromorphic on the punctured plane, plays an important role in this section. First note that if $\Omega \neq C$ and $a \in \partial\Omega$, then $h(z) = f(z - a)$ is meromorphic on Ω and $\|h\|_{QN(\Omega)} \leq K$ since $\delta_\Omega(z) \leq |z - a|$ gives $\delta_\Omega(z) h^\sharp(z) \leq |z - a| f^\sharp(z - a) \leq K$. On the other hand, we can get a lower bound for $\|h\|_{N(A)}$ for certain annuli. Suppose $A = \{z : r^{-1} < |z| < r\}$ and $a = r^{-1}$. Then $\|h\|_{N(A)} \geq h^\sharp(1)/\lambda_A(1)$. But

$$\lambda_A(z) = \frac{\pi}{4 \log r} \frac{1}{|z| \cos\left(\frac{\pi \log |z|}{2 \log r}\right)}$$

so that

$$\lambda_A(1) = \frac{\pi}{4 \log r} = \frac{1}{4m},$$

where $m = \operatorname{mod}(A)$ is the modulus of A. Also

$$h^\sharp(1) = f^\sharp(1 - r^{-1}) \geq \frac{1}{K(1 - r^{-1})} = \frac{1}{K(1 - e^{-\pi m})},$$

so

$$\|h\|_{N(A)} \geq \frac{4m}{K(1 - e^{-\pi m})} \geq Bm$$

for some absolute constant B provided $m \leq 1$.

Now, we introduce various domain constants. Let $m(\Omega) = \sup\{\operatorname{mod}(A) : A$ is an annulus in Ω that separates $\partial\Omega\}$. Here $\partial\Omega$ denotes the boundary of Ω as a subset of the Riemann sphere and $\operatorname{mod}(A) = (1/2\pi) \log(r_2/r_1)$ if $A = \{z : r_1 < |z - a| < r_2\}$. A region Ω is called *uniformly perfect* if $m(\Omega) < \infty$. Define $c(\Omega) = \inf\{\lambda_\Omega(z)\delta_\Omega(z) : z \in \Omega\}$. Beardon and Pommerenke [2] proved that $c(\Omega) > 0$ if and only if $m(\Omega) < \infty$. Finally, let $n(\Omega)$ denote the smallest constant such that $\|f\|_{N(\Omega)} \leq n(\Omega)\|f\|_{QN(\Omega)}$ for all $f \in N(\Omega)$. Recall that $\|f\|_{QN(\Omega)} \leq \|f\|_{N(\Omega)}$ so $1 \leq n(\Omega)$ and $n(\Omega)$ is finite precisely when the two norms are equivalent.

THEOREM 4. *Let Ω be a hyperbolic region in C. $n(\Omega)$ is finite if and only if Ω is uniformly perfect.*

PROOF: First, assume that Ω is uniformly perfect. Then $c = c(\Omega) > 0$ and $\delta_\Omega \geq c/\lambda_\Omega$ yields $c\|f\|_{N(\Omega)} \leq \|f\|_{QN(\Omega)}$, or $n(\Omega) \leq 1/c$.

In order to establish the necessity, we show that there is an absolute constant $D \geq 1$ such that $m(\Omega) \leq Dn(\Omega)$. It is sufficient to show that $m = \mathrm{mod}(A) \leq Dn(\Omega)$ for any annulus A in Ω that separates $\partial\Omega$. This inequality is trivial if $m \leq 1$ since $n(\Omega) \geq 1$. Thus, we assume $m > 1$. There is no harm in assuming that the inner boundary of A meets $\partial\Omega$ in a point a, since shrinking the inner boundary increases the modulus. There is no harm in assuming that $A = \{z : r^{-1} < |z| < r\}$ and $a = r^{-1}$ because $n(\Omega)$ and the modulus of an annulus are both invariant under similarities. Let $h(z) = f(z - a)$ as at the beginning of this section. Then $h \in QN(\Omega)$ and $\|h\|_{QN(\Omega)} \leq K$. From $A \subset \Omega$, we deduce $\lambda_A \geq \lambda_\Omega$ and so $\|h\|_{N(\Omega)} \geq \|h\|_{N(A)} \geq Bm$. Thus,

$$Bm \leq \|h\|_{N(\Omega)} \leq n(\Omega)\|h\|_{QN(\Omega)} \leq n(\Omega)K,$$

or $m \leq Dn(\Omega)$ for some absolute constant $D \geq 1$.

REMARK: The preceding proof and ([1],[2]) show that the quantities $m(\Omega)$, $n(\Omega)$ and $1/c(\Omega)$ are all comparable.

Finally, we observe that there is a simple connection between both normality and quasi-normality and the concept of *uniform normality*. Let Ω be a hyperbolic region in C. It is elementary to show that a meromorphic function f on Ω is quasi-normal on Ω if and only if f is uniformly normal on disks in Ω. This means that there is a constant M such that $\|f\|_{N(\Delta)} \leq M$ for any disk $\Delta \subset \Omega$. Similarly, necessary and sufficient for f to be normal on Ω is that f be uniformly normal on disks and annuli in Ω. This follows from a localization principle of Beardon and Gehring [1, Lemma 2].

Department of Mathematical Sciences, University of Cincinnati, Cincinnati, OH 45221-0025

REFERENCES

1. Beardon, A.F. and Gehring, F.W., *Schwarzian derivatives, the Poincaré metric and the kernel function*, Comment. Math. Helv. **55** (1980), 50–64.
2. Beardon, A.F. and Pommerenke, Ch., *The Poincaré metric of plane domains*, J. London Math. Soc. (2) **18** (1978), 475–483.
3. Campbell, D.M. and Wickes, G., *Characterizations of normal meromorphic functions*, Complex Analysis, Joensuu 1978, pp. 55–72, Lecture Notes in Math. 747, Springer–Verlag, Berlin, 1979.
4. Conway, J.B., "Functions of One Complex Variable," 2nd ed., Springer–Verlag, New York, 1978.
5. Gehring, F.W. and Osgood, B., *Uniform domains and the quasihyperbolic metric*, J. Analyse Math. **36** (1979), 50–74.
6. Gehring, F.W. and Palka, B., *Quasiconformally homogeneous domains*, J. Analyse Math. **30** (1976), 172–199.
7. Kra, I., "Automorphic Functions and Kleinian Groups," W.A. Benjamin, Reading, Mass., 1972.
8. Lehto, O. and Virtanen, K.I., *Boundary behaviour and normal meromorphic functions*, Acta Math **97** (1957), 47–65.
9. Lohwater, A.J. and Pommerenke, Ch., *On normal meromorphic functions*, Ann. Acad. Sci. Fenn. Ser. A.I., No. 550 (1973), 12 pp.
10. Minda, C.D., *Bloch constants*, J. Analyse Math. **41** (1982), 54–84.
11. Pommerenke, Ch., *Uniformly perfect sets and the Poincaré metric*, Arch. Math. **32** (1979), 192–199.
12. Timoney, R.M., *A necessary and sufficient condition for Bloch functions*, Proc. Amer. Math. Soc. **71** (1978), 263–266.

A halfplane version of a theorem of Borel

BY JOHN ROSSI

1. Introduction.

A classical theorem of Borel [2] (c.f. [6, Chapter 5]) states

THEOREM A. *Let $S = \{f_1, \ldots, f_{n+1}\}$ be a set of zero free entire functions such that any proper subset of S is linearly independent. Then S is linearly independent.*

The reader may verify the equivalence of the little Picard theorem to the $n = 2$ case in Theorem A (c.f. [6, p. 112]). In [6, Chapter 5], Nevanlinna extended Theorem A. One may infer from his arguments

THEOREM B. *Let $S = \{f_1, \ldots, f_{n+1}\}$ be a set of meromorphic functions such that any proper subset of S is linearly independent. If S is linearly dependent then*

$$(1.1) \qquad T(r) = O\{\sum_{k=1}^{n+1} [N(r, 1/f_k) + N(r, f_k)] + \log T(r) + \log r\} n.e.$$

where

$$T(r) = \max_{i,j} T(r, f_i/f_j) \quad (i, j = 1, 2, \ldots, n+1).$$

Consequently if

$$(1.2) \qquad N(r, 1/f_k) + N(r, f_k) = o(T(r)) \quad (k = 1, 2, \ldots, n+1),$$

then

$$(1.3) \qquad T(r) = 0(\log r)$$

and hence f_i/f_j is rational for all $i, j = 1, 2, \ldots, n+1$.

We assume familiarity with the Nevanlinna functionals and remind the reader that n.e. means *except possibly on a set of finite measure*.

In short, Theorem B says that if f is a nontrivial linear combination of n linearly independent meromorphic functions $f_1, \ldots, f_n$ with few zeros and poles, then either f has plenty of zeros or f is a rational function times f_i, for some i.

In conversation Simon Hellerstein remarked that some sort of halfplane version of Theorem B should be true. That is, if f is again a nontrivial linear combination of $f_1, \ldots, f_n$ and each f_i has few *nonreal* zeros and poles then either f has plenty of nonreal zeros or else the ratio of any two of the functions involved grows slowly in some sense. (An example which displays the second alternative is $f_1(z) = \sin z$, $f_2(z) = \cos z$, $f = f_1 + f_2$.)

In this note we obtain such a result and deduce as an application

THEOREM 1. *Let R and S be rational, P and Q entire. If*

$$F = R \sin P + S \cos Q$$

has only finitely many nonreal zeros, then Q and P are linear unless F is a rational function times $\exp(\pm iP)$.

We note that the second possibility occurs if and only if $Q \pm P \equiv \text{const.} = c$ and either $\dfrac{R}{2i} + \dfrac{S}{2} \exp(\mp ic)$ or $\dfrac{R}{2i} - \dfrac{S}{2} \exp(\pm ic)$ is identically zero, or P and Q are both constants.

If $S \equiv 0$ and $P(z) = a_n z^n + \cdots + a_0$ is a polynomial, it is easy to see that the zeros of F are "close" to those of $\sin a_n z^n$ and so the theorem is elementary. Similarly we can provide a fairly elementary proof when $Q(z) = P(z) = a_n z^n + \cdots + a_0$. For P, Q arbitrary we know of no elementary proof of Theorem 1. In the special case $S \equiv 0$, $(2\pi)^{-1}P$ has only finitely many nonreal zeros and ones. Theorem 1 then follows from a result of Edrei [**3**, Corollary, p. 227].

2. The Tsuji Characteristic.

In [**8**] (c.f. [**4**] and [**5**]) M. Tsuji introduced a characteristic for functions f meromorphic in the upper halfplane based on the following Jensen-type formula:

$$
\begin{aligned}
(2.1) \quad &\int_1^r \frac{n_0(t,0)}{t^2}\,dt - \int_1^r \frac{n_0(t,\infty)}{t^2}\,dt \\
&= \frac{1}{2\pi} \int_{\sin^{-1}(r^{-1})}^{\pi - \sin^{-1}(r^{-1})} \log |f(re^{i\theta} \sin\theta)| \frac{d\theta}{r \sin^2\theta} + 0(1)
\end{aligned}
$$

Here $n_0(t,0)$ $(n_0(t,\infty))$ denotes the number of zeros (poles) of f in $\{z : |z - \frac{it}{2}| \le \frac{t}{2}, |z| \ge 1\}$.

He defined

$$(2.2) \quad m_0(r,\infty) = m_0(r,f) = \frac{1}{2\pi} \int_{\sin^{-1}(r^{-1})}^{\pi - \sin^{-1}(r^{-1})} \log^+ |f(re^{i\theta} \sin\theta)| \frac{d\theta}{r \sin^2\theta},$$

113

$$(2.3) \qquad m_0(r,a) = m_0(r, 1/(f-a)), \quad a \in \mathbb{C},$$

$$(2.4) \quad N_0(r,\infty) = N_0(r,f) = \int_i^r \frac{n_0(t,\infty)}{t^2}\,dt = \sum_{1 \le r_k < r \sin \phi_K} \left[\frac{\sin \phi_k}{r_k} - \frac{1}{r} \right]$$

where $r_k e^{i\phi_k}$ are the poles of f in $\mathrm{Im}\, z > 0$,

$$(2.5) \qquad N_0(r,a) = N_0(r, 1/(f-a)), \quad a \in \mathbb{C},$$

and

$$(2.6) \qquad T_0(r,f) = m_0(r,f) + N_0(r,f)$$

For f meromorphic in $\mathrm{Im}\, z > 0$, Tsuji proved the following properties:

(A) $m_0(r,a) + N_0(r,a) = T_0(r,f) + 0(1)$, $a \in \mathbb{C}$

(B) If f is also meromorphic in a neighborhood of the origin
$$m_0(r, f'/f) = 0(\log T_0(r,f) + \log r)\, n.e.$$

(C) $T_0(r,f)$ is a monotone increasing function of r.

We also need

(D) $\int_R^\infty \dfrac{m_{0,\pi}(r,f)}{r^3}\,dr \le \int_R^\infty \dfrac{m_0(r,f)}{r^2}\,dr$

where $m_{0,\pi}(r,f) = \dfrac{1}{2\pi} \int_0^\pi \log^+ |f(re^{i\theta})|\,d\theta$. This is proved in [5, pp. 332–333].

Properties (A) and (B) are analogues of Nevanlinna's first fundamental theorem and the lemma of the logarithmic derivative respectively. In [5] it is stated without proof that the requirement in (B) that f be meromorphic near the origin is inessential. Also as in standard Nevanlinna theory (B) implies that for $k = 1, 2, \ldots$

$$(2.7) \qquad m_0(r, f^{(k)}/f) = 0(\log T_0(r,f) + \log r)\, n.e.$$

At first glance the reader may find the Tsuji halfplane theory cumbersome, even unnatural. However, any value distribution theory will require some exhaustion of the halfplane and a compatible Poisson–Jensen formula like (2.1). Properties (A), (B) and (C) are fundamental for Nevanlinna's theory in the plane. The Tsuji theory is based on an exhaustion of the halfplane by circles tangent to the real axis at the origin. The more natural exhaustion by semicircles leads to a theory developed by Nevanlinna and discussed in [4] and [5]. Here (B) holds only for functions meromorphic in $\mathbb{C}$ of restricted growth, namely $\int_0^\infty \dfrac{\log T(r,f)}{r^2}\,dr < \infty$. With this restriction Theorems 1 and 2 also hold in Nevanlinna's halfplane theory.

3. Borel's Theorem in the halfplane.

Our extension of Theorem B involves the Tsuji characteristic.

THEOREM 2. *Let $S = \{f_1, \ldots, f_{n+1}\}$ be a set of functions meromorphic in Im $z > 0$, such that any proper subset of S is linearly independent. If S is linearly dependent then*

$$(3.1) \quad T_0(r) = 0 \left\{ \sum_{k=1}^{n+1} [N_0(r, 1/f_k) + N_0(r, f_k)] + \log T_0(r) + \log r \right\} n.e.$$

where $T_0(r) = \max_{i,j} T_0(r, f_i/f_j)$ $(i, j = 1, \ldots, n+1)$. Further if

$$(3.2) \qquad N_0(r, 1/f_k) + N_0(r, f_k) = o(T_0(r)) \quad (k = 1, \ldots, n+1)$$

then

$$(3.3) \qquad\qquad\qquad T_0(r) = 0(\log r).$$

PROOF: We mimic the method used in [6, Chapter 5]. We assume with no loss of generality that

$$(3.4) \qquad\qquad\qquad f_1 + \cdots + f_{n+1} \equiv 0$$

and set $\phi_k = -f_k/f_{n+1}$ $(k = 1, \ldots, n)$. Thus $\phi_1, \ldots, \phi_n$ are linearly independent and satisfy

$$(3.5) \qquad\qquad\qquad \phi_1 + \cdots + \phi_n = 1.$$

Also

$$(3.6) \qquad\qquad \phi_1^{(i)} + \cdots + \phi_n^{(i)} = 0 \quad (i = 1, \ldots, n-1).$$

Set

$$W = \begin{vmatrix} \phi_1 & \cdots & \phi_n \\ \phi_1' & \cdots & \phi_n' \\ \vdots & & \vdots \\ \phi_1^{(n-1)} & \cdots & \phi_n^{(n-1)} \end{vmatrix}$$

Since $\phi_1, \ldots, \phi_n$ are linearly independent and meromorphic, $W \not\equiv 0$. (The proof of this is a nice little exercise (c.f. [7, Problem 60, p. 108]). Of course it depends crucially on the analytic character of the ϕ's.)

Using (3.5) and (3.6) one can verify that

$$(3.7) \qquad \phi_1 = \frac{W}{\phi_2 \dots \phi_n} \frac{\phi_1 \dots \phi_n}{W} = \frac{\Delta_1}{\Delta}$$

where

$$\Delta = \begin{vmatrix} 1 & \dots & 1 \\ \phi_1'/\phi_1 & \dots & \phi_n'/\phi_n \\ \vdots & & \\ \phi_1^{(n-1)}/\phi_1 & \dots & \phi_n^{(n-1)}/\phi_n \end{vmatrix}$$

and Δ_1 is the minor of Δ with respect to the 1 in the first row, first column (c.f. [1]). By (3.7) and (A)

$$(3.8) \qquad \begin{aligned} m_0(r, \phi_1) &\leq m_0(r, \Delta_1) + m_0(r, 1/\Delta) \\ &= m_0(r, \Delta_1) + m_0(r, \Delta) + N_0(r, \Delta) - N_0(r, 1/\Delta) + 0(1). \end{aligned}$$

Since $\Delta = W/\phi_1 \dots \phi_n$,

$$(3.9) \qquad \begin{aligned} N_0(r, \Delta) - N_0(r, 1/\Delta) &= N_0(r, W) - N_0(r, 1/W) \\ &\quad - \sum_{k=1}^{n} N_0(r, \phi_k) + \sum_{k=1}^{n} N_0(r, 1/\phi_k). \end{aligned}$$

Also (B), (2.7) and the definition of $T_0(r)$ give

$$(3.10) \qquad m_0(r, \Delta) + m_0(r, \Delta_1) = 0(\log T_0(r) + \log r) n.e.$$

Thus (3.8)–(3.10) yield that

$$\begin{aligned} T_0(r, \phi_1) &= m_0(r, \phi_1) + N_0(r, \phi_1) \\ &\leq \sum_{k=1}^{n} N_0(r, 1/\phi_k) + N_0(r, W) + 0(\log T_0(r) + \log r) n.e. \end{aligned}$$

which implies that

$$(3.11) \qquad \begin{aligned} T_0(r, \phi_1) &= 0\{ \sum_{k=1}^{n} [N_0(r, 1/\phi_k) + N_0(r, \phi_k)] \\ &\quad + 0(\log T_0(r) + \log r)\} n.e. \end{aligned}$$

Clearly (3.11) is also true for ϕ_i, $i = 2, \dots, n$. In fact (3.11) is true with ϕ_1 replaced by any ratio of the form f_i/f_j. Thus (3.11) implies (3.1).

If we assume (3.2), then (3.11) gives that $T_0(r) = o(T_0(r)) + 0(\log r) n.e.$ which is impossible unless

$$(3.12) \qquad T_0(r) = 0(\log r) n.e.$$

Since (C) implies that $T_0(r)$ is monotone increasing, (3.12) implies (3.3). The proof of Theorem 2 is complete.

4. Proof of Theorem 1.

We first assume that

$$(4.1) \qquad R \not\equiv 0, \quad S \not\equiv 0, \quad P \pm Q \not\equiv \text{const.}, \quad P \not\equiv \text{const.}, \quad Q \not\equiv \text{const.}$$

and write

$$F = \frac{R}{2i} e^{iP} + \frac{-R}{2i} e^{-iP} + \frac{S}{2} e^{iQ} + \frac{S}{2} e^{-iQ}$$

$$= f_1 + f_2 + f_3 + f_4 \text{ (say)}$$

Then $\{F, f_1, f_2, f_3, f_4\}$ satisfies the independence criteria of Theorem 2. Since R, S are rational, $N_0(r, R)$, $N_0(r, 1/R)$, $N_0(r, S)$ and $N_0(r, 1/S)$ are all bounded. If $T_0(r) \neq 0(\log r)$, the monotonicity of $T_0(r)$ shows that $T_0(r) \to \infty$ and so (3.2) holds. But then all the hypotheses of Theorem 2 are met and $T_0(r) = 0(\log r)$, a contradiction. Thus

$$(4.2) \qquad T_0(r) = 0(\log r)$$

If we define $\overline{T}_0(r, f)$ to be the Tsuji characteristic for functions meromorphic in the lower halfplane and define $\overline{T}_0(r)$ in the obvious way we get as in (4.2) that

$$(4.3) \qquad \overline{T}_0(r) = 0(\log r).$$

Clearly (4.2) and (4.3) imply

$$(4.4) \qquad \begin{aligned} m_0(r, f_1/f_2) &= 0(\log r) \\ \overline{m}_0(r, f_1/f_2) &= 0(\log r) \end{aligned}$$

where $\overline{m}_0$ is the lower halfplane analogue of m_0. Since $f_1/f_2 = -e^{2iP}$ is entire we have

$$(4.5) \qquad T(r, f_1/f_2) = m(r, f_1/f_2) = m_{0,\pi}(r, f_1/f_2) + m_{\pi, 2\pi}(r, f_1/f_2)$$

where T is the usual Nevanlinna characteristic and $m_{\pi, 2\pi}$ is defined in the obvious way. By (4.4), (4.5), (D) and its lower halfplane analogue it follows that

$$(4.6) \qquad \int_R^\infty \frac{T(r, f_1/f_2)}{r^3} dr \leq \int_R^\infty \frac{0(\log r)}{r^2} dr = 0 \frac{(\log R)}{R}$$

$$117$$

Since T is increasing

$$(4.7) \qquad\qquad T(R, f_1/f_2) = 0(R \log R).$$

Hence the order of e^{2iP} equals one. This means that P is linear. Similarly, analysis of f_3/f_4 gives that Q is linear.

It remains to drop the assumptions (4.1). Say $Q + P \equiv c$ where c is a constant. We may assume that F is not a rational function times $\exp(\pm iP)$; otherwise we have nothing to prove. Hence F can be written as the sum of two linearly independent functions f_1 and f_2 where $f_1 = \left[\dfrac{R}{2i} + \exp(-ic)\dfrac{S}{2} \right] e^{iP}$ and $f_2 = \left[\dfrac{-R}{2i} + \exp(+ic)\dfrac{S}{2} \right] e^{-iP}$. (See the paragraph after the statement of Theorem 1.) The only difference from before is that f_1/f_2 need not be entire. But clearly since R and S are rational, and c is a constant, we obtain $T(r, f_1/f_2) = m(r, f_1/f_2) + 0(\log r)$ which does not affect the original argument. The other assumptions in (4.1) can be dealt with similarly. The proof of Theorem 1 is complete.

Acknowledgement.

We would like to thank Simon Hellerstein for some extremely useful conversations and the referee for carefully reading the manuscript and offering valuable suggestions to improve it.

Department of Mathematics, Virginia Polytechnic Institute and State University, Blacksburg, VA 24061

REFERENCES

1. Bloch, A., *Sur les systems de fonctions holomorphes a varietes lineaires lacunaires*, Ann. de l'Ec. Norm. **42** (1925).
2. Borel, E., "Lecons sur les fonctions entieres," Gauthier–Villars, Paris, 1921.
3. Edrei, A., *Meromorphic functions with three radially distributed values*, Trans. Amer. Math. Soc. **78** (1955), 276–294.
4. Hellerstein, S. and Yang, C.C., *Half-plane Tumura–Clunie theorems and the real zeros of successive derivatives*, J. London Math. Soc. **4** (1972), 469–481.
5. Levin, B. Ja. and Ostrovskii, I.V., *The dependence of the growth of an entire function on the distribution of the zeros of its derivatives*, Amer. Math. Soc. Transl. (2) **32** (1963), 322–357.
6. Nevanlinna, R., "Le theoreme de Picard–Borel et la theorie des fonctions meromorphes," Gauthier–Villars, 1929.
7. Polya, G. and Szegö, G., "Problems and Theorems in Analysis – Volume II," Springer–Verlag, New York, 1976.
8. Tsuji, M., *On Borel's directions of meromorphic functions of finite order. I*, Tohuku Math. J. **2** (1950), 97–112.

An index theorem on singular points and cusps of quadrature domains

BY MAKOTO SAKAI

Quadrature domains were investigated by P.J. Davis [3] and D. Aharonov–H.S. Shapiro [1]. Generalized quadrature domains were defined and studied by B. Gustafsson [4] and others, see, for example, [5].

In this paper, we return to original quadrature domains, namely, quadrature domains of point differential functionals of finite order and show an index theorem. We shall give its application and discuss the case of order two.

Let D be a bounded domain of the complex z-plane and let $AL^1(D)$ be the class of functions analytic and integrable in D. Let

$$L(f) = \sum_{j=1}^{q} \sum_{k=0}^{n_j-1} a_{jk} f^{(k)}(z_j)$$

be a point differential functional of finite order defined for functions analytic in a neighborhood of $\cup_{j=1}^{q}\{z_j\}$, where z_j are distinct points in D and a_{jk} are constants independent of f. We assume that $a_{jn_j-1} \neq 0$ for every j. We put $p = \sum_{j=1}^{q} n_j$ and call it the order of L.

We call D a quadrature domain of L if

$$\int_D f(z)\,dxdy = L(f) \qquad (z = x + iy)$$

for every f in $AL^1(D)$.

A simple example of a quadrature domain is a disk: Let $L(f) = a_{10}f(z_1)$ and $a_{10} > 0$. The quadrature domain of L is a disk of radius $(a_{10}/\pi)^{1/2}$ and centered at z_1. This is the case of order one. If the order is greater than one, it is not easy to determine all quadrature domains of the functional. In 1976, D. Aharonov–H.S. Shapiro [1] proved that every quadrature domain is finitely connected and each its boundary component is a point or an algebraic curve which may have a finite number of cusps pointing into the quadrature domain.

If D has a boundary component which reduces to a point ς, then $D \cup \{\varsigma\}$ is also a quadrature domain of L. Therefore, throughout this paper, we assume that each boundary component is an algebraic curve. We denote by b the number of boundary components of D.

Let $\hat{D}(\varsigma) = \int_D 1/(z - \varsigma)dxdy$ and $\hat{L}(\varsigma) = L(1/(z - \varsigma))$. Since $1/(z - \varsigma) \in AL^1(D)$ for $\varsigma \notin D$, $\hat{D}(\varsigma) = \hat{L}(\varsigma)$ outside of D. Since $\partial\hat{D}/\partial\bar{z} = -\pi$ in D, $\hat{D}(z) + \pi\bar{z}$ is a function, say $A(z)$, analytic in D and continuous on the closure of D. We set

$$S(z) = \{A(z) - \hat{L}(z)\}/\pi.$$

This function is meromorphic in D and has p poles in D, namely, z_j is a pole of order n_j for every j. It is continuous and satisfies $S(z) = \bar{z}$ on the boundary ∂D of D. We call it the Schwarz function of ∂D.

Next we shall define a singular point of D by using S. Set $v(z) = S(z) - \bar{z}$. It vanishes on ∂D and vanishes at only a finite number of points in D. We call the points in D singular points of D and denote by E the set of singular points of D, namely, $E = \{\varsigma \in D;\ S(\varsigma) = \bar{\varsigma}\}$. We note that $S(\varsigma) = \bar{\varsigma}$ if and only if $\hat{D}(\varsigma) = \hat{L}(\varsigma)$.

We denote by $\iota(\varsigma)$ the index of v at $\varsigma \in E$. It is defined as the degree of a mapping $z \mapsto v(z)/|v(z)|$, where z varies on a small circle centered at ς. We shall show in Section 1 that $\iota(\varsigma) = -1,\ 0$ or $+1$. Let us denote by s_{-1}, s_0 and s_{+1} the number of singular points in E with index $-1,\ 0$ and $+1$, respectively.

Let ς be a cusp on ∂D and let $s(\varsigma)$ denote the unit vector at ς tangent to ∂D and pointing into D. Let Γ_ε be a half circle defined by $\Gamma_\varepsilon = \{z;\ |z-\varsigma| = \varepsilon$ and $|\arg(z - \varsigma) - \arg s(\varsigma)| \leq \pi/2\}$. Γ_ε is oriented counterclockwise. In Section 1, we shall show that $\int_{\Gamma_\varepsilon} d\arg v(z) = -\pi + 0(\varepsilon)$ or $\pi + 0(\varepsilon)$. We define the index $\iota(\varsigma)$ of the cusp ς by $-1/2$ or $+1/2$, respectively. We denote by $c_{-1/2}$ and $c_{+1/2}$ the number of cusps on ∂D with index $-1/2$ and $+1/2$, respectively.

Our purpose is to show the following index theorem:

THEOREM. *It follows that*

$$s_{+1} - s_{-1} + c_{+1/2} = p + b - 2.$$

This paper consists of five sections. The main theorem will be proved in Section 1. We shall construct examples of singular points and cusps in Sections 2 and 3. We discuss quadrature domains for subharmonic functions in Section 4. We give two applications of the index theorem in Section 5.

The author wishes to express his heartfelt gratitude to Professor H.S. Shapiro. Professor Shapiro discussed the present material with him at the University of Maryland in 1985 and suggested our index theorem. He is grateful to B. Gustafsson for pointing out mistakes in the first draft and to the referee for helpful comments.

§1. Proof of Theorem.

First we shall show that $\iota(\varsigma) = -1, 0$ or $+1$ for ς in E. In a neighborhood of ς, v has an expansion

$$
\begin{aligned}
v(\varsigma + \tau) &= S(\varsigma + \tau) - \overline{\varsigma + \tau} \\
&= \{S(\varsigma) + S'(\varsigma)\tau + \ldots\} - \overline{\varsigma + \tau} \\
&= \{S'(\varsigma)\tau - \overline{\tau}\} + 0(\tau^2).
\end{aligned}
$$

If $|S'(\varsigma)| > 1$ (resp. $|S'(\varsigma)| < 1$), then v is nondegenerate at ς and $\iota(\varsigma) = 1$ (resp. $\iota(\varsigma) = -1$). If $|S'(\varsigma)| = 1$, v is degenerate at ς. We set $S'(\varsigma) = e^{i\phi}$ and $t = e^{i\phi/2}\tau = re^{i\theta}$. Then

$$
\begin{aligned}
v(\varsigma + \tau) &= e^{i\phi/2}\{t - \overline{t} + 0(t^2)\} \\
&= re^{i\phi/2}\{(2\sin\theta)i + 0(r)\}.
\end{aligned}
$$

Since $(\partial/\partial\theta)\mathrm{Im}\{(2\sin\theta)i + 0(r)\} = 2\cos\theta + 0(r)$, $\mathrm{Im}\{(2\sin\theta)i + 0(r)\} = 0$ has a unique solution θ_1 as θ increases from $-\pi/2$ to $\pi/2$ for sufficiently small fixed $r > 0$. Since $v \neq 0$ for small $r > 0$, $(2\sin\theta_1)i + 0(r)$ is positive or negative and $\arg v$ increases the amount $\pi + 0(r)$ or $-\pi + 0(r)$, respectively. The same holds when θ increases from $\pi/2$ to $3\pi/2$ and so $\arg v$ varies totally -2π, 0 or $+2\pi$ as θ varies counterclockwise from $-\pi/2$ to $3\pi/2$. Thus $\iota(\varsigma) = -1, 0$ or $+1$.

Next we shall show that $\int_{\Gamma_e} d\arg v(z) = \pm\pi + 0(\varepsilon)$ for a cusp ς on ∂D. Let $H = \{t \in \mathbb{C}; |t| < 1 \text{ and } \mathrm{Re}\, t > 0\}$ and take a neighborhood V of ς so that $D \cap V$ is a Jordan domain. Let $z = F(t)$ be a one-to-one conformal mapping of H onto $D \cap V$. The mapping F can be extended to a homeomorphism of the closure of H onto that of $D \cap V$. We denote it again by F. We may assume that $F(0) = \varsigma$ and $F(I) \subset \partial D$, where $I = \{iy; -1 < y < 1\}$. Set

$$
\tilde{F}(t) = \begin{cases} F(t) & t \in H \cup I \\ \overline{S(F(t^*))} & t \in H^*, \end{cases}
$$

where $t^* = -\bar{t}$ is the reflection of t with respect to the imaginary axis and $H^* = \{t^*;\ t \in H\}$.

Since $F(I) \subset \partial D$ and $\overline{S(z)} = z$ on ∂D, $\tilde{F}$ is continuous on the unit disk $H \cup I \cup H^*$ and so it is an analytic continuation of F. We denote it again by F. Let $F(t) = \varsigma + s(\varsigma) \sum_{n=1}^{\infty} a_n t^n$ be an expansion of F at the origin. Since ς is a cusp on ∂D, $a_1 = 0$. By substituting iy for t, where y is a small positive number, we obtain $a_2 > 0$.

For t in H, t^* belongs to H^* and $F(t^*) = \overline{S(F(t^{**}))} = \overline{S(F(t))} = \overline{S(z)}$. Hence

$$S(z) = \overline{F(t^*)} = \overline{F(-\bar{t})} = \bar{\varsigma} + \overline{s(\varsigma)} \sum_{n=1}^{\infty} (-1)^n \bar{a}_n t^n$$

and

$$v(z) = S(z) - \bar{z} = \overline{s(\varsigma)} \sum_{n=2}^{\infty} \bar{a}_n \{(-1)^n t^n - \bar{t}^n\}.$$

Putting $t = re^{i\theta}$, we obtain

$$v(z) = a_2 r^2 \overline{s(\varsigma)} \{(2\sin 2\theta)i + 0(r)\}$$

in a neighborhood of ς. Since $(\partial/\partial\theta)\mathrm{Im}\{(2\sin 2\theta)i + 0(r)\} = 4\cos 2\theta + 0(r)$, $\mathrm{Im}\{(2\sin 2\theta)i + 0(r)\} = 0$ has a unique solution θ_0 as θ increases from $-\pi/4$ to $\pi/4$ for sufficiently small fixed $r > 0$. Hence $\int_{\Gamma_e} d\arg v(z) = \pi + 0(\varepsilon)$ if $(2\sin 2\theta_0)i + 0(r)$ is positive and $\int_{\Gamma_e} d\arg v(z) = -\pi + 0(\varepsilon)$ if $(2\sin 2\theta_0)i + 0(r)$ is negative.

Now we shall show our main theorem. Let P be the set of poles of S and set $D_\varepsilon = \{z \in D;\ \mathrm{dist}(z, P \cup \partial D) > \varepsilon\}$ for $\varepsilon > 0$. For sufficiently small ε, D_ε is connected and all singular points of v in D are contained in D_ε, namely, $E \subset D_\varepsilon$.

The Poincaré theorem asserts that

$$\sum_{\varsigma \in E} \iota(\varsigma) = \text{the rotation index of } v \text{ on } \partial D_\varepsilon,$$

where D_ε is oriented positively in the usual sense. We shall compute the rotation index of v on ∂D_ε. If z varies on the boundary of small disks centered at poles of S, $v(z) = S(z) - \bar{z}$ varies like $S(z)$ does and so the rotation index of v on there is equal to p, the number of poles of S.

To compute the rotation index of v on the other boundary components of D_ε, we treat first the case that there are no cusps on ∂D. Let z be a point on the outer boundary of D_ε. For small $\varepsilon > 0$, we can find the unique closest point ς on ∂D to z in ∂D and z can be expressed as $z = \varsigma + i\varepsilon s(\varsigma)$, where $s(\varsigma)$ denotes the unit vector at ς tangent positively to ∂D. Since $\partial f/\partial s = (\partial f/\partial \varsigma)(\partial \varsigma/\partial s) + (\partial f/\partial \overline{\varsigma})(\partial \overline{\varsigma}/\partial s) = f'(\varsigma)s(\varsigma)$ for every analytic function f in a neighborhood of ς, where s denotes the length parameter on ∂D,

$$S'(\varsigma) = \frac{1}{s(\varsigma)}\frac{\partial S}{\partial s} = \frac{1}{s(\varsigma)}\frac{\partial \overline{\varsigma}}{\partial s} = \frac{1}{s(\varsigma)}\overline{s(\varsigma)}.$$

Hence

$$\begin{aligned}
v(z) &= S(\varsigma + i\varepsilon s(\varsigma)) - \overline{\varsigma + i\varepsilon s(\varsigma)} \\
&= S(\varsigma) + S'(\varsigma)i\varepsilon s(\varsigma) + 0(\varepsilon^2) - \overline{\varsigma} + i\varepsilon\overline{s(\varsigma)} \\
&= 2i\varepsilon\overline{s(\varsigma)} + 0(\varepsilon^2)
\end{aligned}$$

and so the rotation index of v on the outer boundary of D_ε is equal to -1. The same argument about the other boundary components shows that the rotation index of v on the remaining boundary components is equal to $b - 1$. Therefore we obtain

$$s_{+1} - s_{-1} = \sum_{\varsigma \in E} \iota(\varsigma) = p - 1 + b - 1 = p + b - 2.$$

If there are cusps on ∂D, it is necessary to argue more precisely. Assume that there is just one cusp, say ς, on ∂D. We divide ∂D_ε into two parts: $-\Gamma_\varepsilon$ and $(\partial D_\varepsilon)\backslash\Gamma_\varepsilon$. On $(\partial D_\varepsilon)\backslash\Gamma_\varepsilon$, $v(z) = 2i\varepsilon\overline{s(\varsigma)} + 0(\varepsilon^2)$ as before, and so the variation of $\arg v$ on $(\partial D_\varepsilon)\backslash\Gamma_\varepsilon$ is equal to $2\pi(p + b - 2) - \pi + 0(\varepsilon)$. On $-\Gamma_\varepsilon$, it is equal to $\pi + 0(\varepsilon)$ (resp. $-\pi + 0(\varepsilon)$) if $\iota(\varsigma) = -1/2$ (resp. $\iota(\varsigma) = +1/2$). Hence the rotation index of v on ∂D_ε is equal to $p + b - 2$ if $\iota(\varsigma) = -1/2$ and $p + b - 3$ if $\iota(\varsigma) = +1/2$. Thus the rotation index of v on ∂D_ε is equal to $p + b - 2 - c_{+1/2}$ in general and the proof is complete.

§2. Singular points with index -1, 0, or $+1$.

Let D be a simply connected domain and let $z = R(w)$ be a one-to-one conformal mapping of $U = \{w; |w| < 1\}$ onto D. P.J. Davis [3, (8.4) on p. 52, Theorems on p. 154 and p. 158] showed that D is a quadrature domain of L of order p if and only if R is a rational function of order p and the Schwarz function S of ∂D is given by

$$S(z) = \overline{R}\left(\frac{1}{R^{-1}(z)}\right),$$

where R^{-1} denotes the inverse function of R defined in D and $\overline{R}$ is a function defined by $\overline{R}(w) = \overline{R(\overline{w})}$.

To construct examples of singular points with index -1, 0 or $+1$, let

$$z = R(w) = \frac{w + a_2 w^2}{1 + b_3 w^3},$$

where $|a_2| < 1/5$ and $0 < |b_3| < 1/5$. If $R(w_1) = R(w_2)$ for w_1 and w_2 in U, then $(w_2 - w_1)\{1 + a_2(w_1 + w_2) - b_3 w_1 w_2(w_1 + w_2) - a_2 b_3 w_1^2 w_2^2\} = 0$. Since $|a_2| < 1/5$, $|b_3| < 1/5$ and $|w_j| < 1$, $|a_2(w_1 + w_2) - b_3 w_1 w_2(w_1 + w_2) - a_2 b_3 w_1^2 w_2^2| < 1$. This implies that $w_1 = w_2$ and so R is univalent in U. It satisfies $z = R(w) = w + a_2 w^2 + 0(w^3)$, $w = R^{-1}(z) = z - a_2 z^2 + 0(z^3)$ and $\overline{R}(1/w) = (w^2 + \overline{a}_2 w)/(w^3 + \overline{b}_3) = (1/\overline{b}_3)(\overline{a}_2 w + w^2 + 0(w^3))$. Hence

$$S(z) = \frac{\overline{a}_2}{\overline{b}_3} z + \frac{1}{\overline{b}_3}(1 - |a_2|^2)z^2 + 0(z^3)$$

in a neighborhood of $z = 0$ and $z = 0$ is a singular point of $D = R(U)$. It follows that $\iota(0) = +1$ if $|a_2| > |b_3|$ and $\iota(0) = -1$ if $|a_2| < |b_3|$.

To construct an example of a singular point with index 0, take $b_3 = a_2 \neq 0$. We shall show that $z = 0$ is the desired one if a_2 is not pure imaginary. Set $(1/\overline{a}_2)(1 - |a_2|^2) = \alpha + i\beta$, $\alpha \neq 0$, and $z = t = re^{i\theta}$. Then, as in Section 1, we obtain

$$v(z) = t - \overline{t} + (\alpha + i\beta)t^2 + 0(t^3)$$

$$= r\{(2\sin\theta)i + (\alpha + i\beta)re^{i2\theta} + 0(r^2)\}.$$

Let $\theta_j = \theta_j(r)$, $j = 1, 2$, be two solutions of $\operatorname{Im}(v/r) = 2\sin\theta + r(\alpha\sin 2\theta + \beta\cos 2\theta) + 0(r^2) = 0$. We may assume that $\theta_1(r) = 0(r)$ and $\theta_2(r) = \pi + 0(r)$. Since $\operatorname{Re}(v/r) = r\{(\alpha\cos 2\theta - \beta\sin 2\theta) + 0(r)\}$, the signs of $v(re^{i\theta_1})$ and $v(re^{i\theta_2})$ are the same with that of $\alpha \neq 0$ for sufficiently small $r > 0$. Hence $\iota(0) = 0$.

§3. **Cusps with index $-1/2$ or $+1/2$.**

Let α be a real number with $0 \leq \alpha \leq 1$ and let

$$z = R(w) = (w + 1)^2 + \alpha(w + 1)^3.$$

We shall show that $z = 0$ is a cusp of $D = R(U)$ with index $+1/2$ if $0 \leq \alpha < 1$ and with index $-1/2$ if $\alpha = 1$.

First we shall see that R is univalent in U. We use a result due to D.A. Brannan [**2**, Theorem 2, (b)]: A polynomial $w + \beta_2 w^2 + \beta_3 w^3$ with real β_2 and β_3 satisfying $0 \leqq \beta_3 \leqq 1/5$ is univalent in U if and only if $|\beta_2| \leqq (1 + 3\beta_3)/2$. Since

$$R(w) = (1 + \alpha) + (2 + 3\alpha)\{w + \frac{1 + 3\alpha}{2 + 3\alpha}w^2 + \frac{\alpha}{2 + 3\alpha}w^3\},$$

$0 \leqq \alpha/(2 + 3\alpha) \leqq 1/5$ if $0 \leqq \alpha \leqq 1$ and

$$\frac{1 + 3\alpha}{2 + 3\alpha} = (1 + 3 \cdot \frac{\alpha}{2 + 3\alpha})/2,$$

R is univalent in U for α with $0 \leqq \alpha \leqq 1$. Let $w = w(t) = (t - 1)/(t + 1)$ be a one-to-one conformal mapping of $\{t;\ \mathrm{Re}\ t > 0\}$ onto U and set $F(t) = R(w(t))$. Then $F(t) = 4\{t^2 - 2(1 - \alpha)t^3 + 3(1 - 2\alpha)t^4 - 4(1 - 3\alpha)t^5 + 0(t^6)\}$ in a neighborhood of $t = 0$. Hence, setting $t = re^{i\theta}$, we obtain

$$\mathrm{Re}\{v(z)/4\} = 4(1 - \alpha)r^3 \cos 3\theta + 8(1 - 3\alpha)r^5 \cos 5\theta + 0(r^7).$$

If $0 \leqq \alpha < 1$, then $\mathrm{Re}\ v(z) > 0$ for $\theta_0 = \theta_0(r)$ with small $r > 0$. If $\alpha = 1$, then $\mathrm{Re}\{v(z)/4\} = -16r^5(\cos 5\theta + 0(r^2))$ and $\mathrm{Re}\ v(z) < 0$ for θ_0. Thus the index of the cusp $z = 0$ is equal to $+1/2$ if $0 \leqq \alpha < 1$ and $-1/2$ if $\alpha = 1$.

§4. Quadrature domains for subharmonic functions.

We call D a quadrature domain of L for subharmonic functions if (i) L is a positive functional, namely, $n_j = 1$ and $a_{j0} > 0$ for every j with $1 \leqq j \leqq q = p$ and (ii) D satisfies

$$(4.1) \qquad \int_D f(z)dxdy \geqq \sum_{j=1}^{p} a_{j0} f(z_j)$$

for every subharmonic and integrable function f in D. If f is harmonic, f and $-f$ are both subharmonic and so equality holds in (4.1). Since $\mathrm{Re}\ f$ and $\mathrm{Im}\ f$ are both harmonic for every analytic function f, D is a quadrature domain of L in the original sense.

In this section, we shall determine the index of a cusp of a quadrature domain for subharmonic functions and that of a singular point ς satisfying

$$(4.2) \qquad \int_D \log |z - \varsigma|dxdy = \sum_{j=1}^{p} a_{j0} \log |z_j - \varsigma|.$$

If ς is not contained in D, then $\log|z - \varsigma|$ is harmonic in D and so (4.2) holds for every ς on the complement of D. Since $\Delta_\varsigma \int_D \log|z - \varsigma| dx dy = 2\pi$ in D, where Δ_ς denotes the Laplacian with respect to ς,

$$\int_D \log|z - \varsigma| dx dy = \frac{\pi}{2}|\varsigma|^2 + h(\varsigma)$$

for a function h harmonic in D and of class C^1 in the whole plane. We set

$$u(\varsigma) = \frac{2}{\pi}\left\{ \sum_{j=1}^{p} a_{j0}\log|z_j - \varsigma| - h(\varsigma) \right\}.$$

Then, by (4.1), $|\varsigma|^2 \geqq u(\varsigma)$ in D and equality holds on the complement of D. Since u is of class C^1 in a neighborhood of ∂D, $\partial|\varsigma|^2/\partial\varsigma = \partial u(\varsigma)/\partial\varsigma$ on ∂D. Hence

$$\bar\varsigma = \frac{\partial u(\varsigma)}{\partial\varsigma} = \frac{2}{\pi}\left\{ \frac{1}{2}\sum_{j=1}^{p}\frac{a_{j0}}{\varsigma - z_j} - \frac{\partial h(\varsigma)}{\partial\varsigma} \right\}$$

on ∂D. The function in the right-hand side of the equation is meromorphic in D and so it is the Schwarz function S of ∂D.

Let ς be a cusp on ∂D. We may assume that $s(\varsigma) = 1$. Let C be a curve starting from ς defined by $C = \{z \in D;\ \mathrm{Im}\ v(z) = 0\}$. Then $\partial(|z|^2 - u(z))/\partial y = 0$ on C, because $\partial(|z|^2 - u(z))/\partial z = \bar z - S(z) = -v(z)$ is real on C. Let $z(s)$ be a point on C, where s denotes the arc length of C from ς to $z(s)$. Then, by the mean value theorem,

$$|z(s)|^2 - u(z(s)) = s\frac{\partial}{\partial s}\{|z(\lambda s)|^2 - u(z(\lambda s))\}$$

for some λ with $0 < \lambda < 1$ and the right-hand side is equal to $s(\partial/\partial x)\{|z(\lambda s)|^2 - u(z(\lambda s))\}(\partial x(\lambda s)/\partial s)$. Since $\partial x(\lambda s)/\partial s > 0$ for small $s > 0$, we obtain

$$-\mathrm{Re}\ v(z(\lambda s)) = \frac{1}{2}\frac{\partial}{\partial x}\{|z(\lambda s)|^2 - u(z(\lambda s))\}$$

$$= \frac{1}{2}\{s(\partial x(\lambda s)/\partial s)\}^{-1}\{|z(s)|^2 - u(z(s))\}$$

$$\geqq 0.$$

Hence ς is a cusp with index $-1/2$.

Let ς be a singular point of D satisfying (4.2). Then, by definition, $|\varsigma|^2 = u(\varsigma)$. We may assume that $S'(\varsigma) \geq 0$. Let C be a curve passing through ς defined by $C = \{z \in D;\ \mathrm{Im}\ v(z) = 0\}$. We divide C into C_+ and C_- defined by $C_+ = C \cap \{z;\ \mathrm{Re}(z-\varsigma) \geq 0\}$ and $C_- = C \cap \{z;\ \mathrm{Re}(z-\varsigma) < 0\}$. By replacing C by C_+ and C_-, and using the same argument as above, we obtain $-\mathrm{Re}\ v(z(\lambda s)) \geq 0$ for $z(s)$ on C_+ and $-\mathrm{Re}\ v(z(\lambda s)) \leq 0$ for $z(s)$ on C_-. Hence ς is a singular point with index -1.

§5. Applications of the index theorem.

We give two applications in this section and study the special cases of quadrature domains: namely, (1) $S'(z) \neq 0$ in D and (2) $p = 2$.

First we prepare

LEMMA. *If $S'(z)$ has m zeros in D, then*

$$m + c = p + q + 2(b - 2),$$

where $c = c_{-1/2} + c_{+1/2}$ denotes the number of cusps on ∂D.

PROOF: Since $S'(z)$ has $p + q$ poles in D, by the Poincaré theorem or the argument principle this time, $m - p - q$ is equal to the rotation index of $S'(z)$ on ∂D. We have seen in Section 1 that $S'(z) = s(z)^{-2}$ on ∂D and so the rotation index of $S'(z)$ on ∂D is equal to $2(b - 2) - c$.

(1)   The case that $S'(z) \neq 0$ in D

If $S'(z) \neq 0$ in D, then $\log |S'(z)|$ is superharmonic in D and $\log |S'(z)| = -2 \log |s(z)| = 0$ on ∂D. Hence $\log |S'(z)| > 0$ in D, namely, $|S'(z)| > 1$ in D and so v is nondegenerate at every singular point and the index is equal to $+1$.

Next we shall show that the index of every cusp is equal to $+1/2$. Let ς be a cusp on ∂D. We may assume that $s(\varsigma) = 1$. Let C be a curve starting from ς and defined by $C = \{z \in D;\ \mathrm{Im}\ v(z) = 0\}$. Let $z = z(s)$ be a point on C and set $S'(z) = \alpha + i\beta$ and $\partial z/\partial s = \xi + i\eta$. Then

$$\frac{\partial v}{\partial s} = \frac{\partial v}{\partial z}\frac{\partial z}{\partial s} + \frac{\partial v}{\partial \bar{z}}\frac{\partial \bar{z}}{\partial s}$$

$$= S'(z)\frac{\partial z}{\partial s} - \frac{\overline{\partial z}}{\partial s}$$

$$= \{(\alpha - 1)\xi - \beta\eta\} + i\{(\alpha + 1)\eta + \beta\xi\}$$

is real on C and so

$$(\alpha + 1)\frac{\partial v}{\partial s} = \{(\alpha^2 - 1)\xi - (\alpha + 1)\beta\eta\} = (\alpha^2 + \beta^2 - 1)\xi.$$

Since

$$S'(z) = \frac{dS}{dt} \frac{1}{\frac{dz}{dt}} = \frac{\sum n(-1)^n \bar{a}_n t^{n-1}}{\sum n a_n t^{n-1}} = 1 + 0(t),$$

$\alpha + 1 > 0$ for small s. Combining $\alpha^2 + \beta^2 = |S'(z)|^2 > 1$ and $\xi = \mathrm{Re}(\partial z/\partial s) = \partial x/\partial s > 0$ on C with the above equation, we obtain $\partial v/\partial s > 0$ on C for small s. By the mean value theorem, $v(z(s)) = s(\partial v(z(\lambda s))/\partial s)$ for some λ with $0 < \lambda < 1$. Hence $\mathrm{Re}\, v(z(s)) = v(z(s)) > 0$ and so ς is a cusp with index $+1/2$.

Now we apply our index theorem. We have proved that $s_{-1} = s_0 = c_{-1/2} = 0$ if $m = 0$ and so $s_{+1} + c_{+1/2} = p + b - 2$ by Theorem and $c_{+1/2} = p + q + 2(b - 2)$ by Lemma. Hence $b + q + s_{+1} = 2$. Since $b \geq 1$ and $q \geq 1$, we get $b = q = 1$, $s_{+1} = 0$ and $c_{+1/2} = p - 1$. Hence, by the result due to P.J. Davis mentioned at the beginning of Section 2, there is a rational function $z = R(w)$ of order p which is analytic and univalent in the unit disk U and $D = R(U)$. Since $q = 1$, we may assume that

$$S(z) = \overline{R}\left(\frac{1}{R^{-1}(z)}\right)$$

has a pole of order p at $R(0)$, namely, $\overline{R}$ has a pole of order p at ∞. Hence R is a polynomial of degree p which is univalent in U. Since $c_{+1/2} = p - 1$, $R'(w)$ has distinct $p - 1$ zeros on ∂U and so R can be expressed in the form

$$(5.1) \qquad R(w) = a(w + a_2 w^2 + \cdots + a_p w^p) + a_0$$

with $|a_p| = 1/p$.

Conversely, if D is the image of U under a polynomial R which is univalent in U and expressed in the form (5.1) with $|a_p| = 1/p$, then $R'(w) = a(1 + 2a_2 w + \cdots + pa_p w^{p-1})$ and $|pa_p| = 1$. Since $R'(w) \neq 0$ in U, all zeros of R' are on ∂U and so $R'(w) \neq 0$ in $\{w;\ |w| > 1\}$. Hence $\overline{R}'(w) \neq 0$ in $\{w;\ |w| > 1\}$ and so

$$S'(z) = \overline{R}'\left(\frac{1}{R^{-1}(z)}\right) \cdot \frac{-1}{R^{-1}(z)^2} \cdot R^{-1'}(z) \neq 0$$

in D.

Thus we have proved that $S'(z) \neq 0$ in D if and only if $D = R(U)$ for a polynomial R which is univalent in U and expressed in the form (5.1) with $|a_p| = 1/p$. In this case, the quadrature domain $R(U)$ has the very nice geometric property. The tangent vector $s(R(e^{i\phi}))$ to the boundary turns at a constant rate as ϕ varies except at the cusps. For details, we refer to T.J. Suffridge [6].

(2) The case $p = 2$

H.S. Shapiro has recently proved that $s + c = 1$ if $p = 2$, where $s = s_{-1} + s_0 + s_{+1}$ and $c = c_{-1/2} + c_{+1/2}$. In what follows, we shall show that $s_{-1} = s_0 = c_{-1/2} = 0$ and $s_{+1} + c_{+1/2} = 1$ if $p = 2$.

First we shall follow the argument of D. Aharonov–H.S. Shapiro [1] and show that $b = 1$ if $p = 2$. Let D be a quadrature domain of L of order p and let E be the set of singular points of D. D. Aharonov–H.S. Shapiro [1] showed that $E \cup \partial D$ is a subset of $\{z = x + iy;\ P(x, y) = 0\}$ for a polynomial P with real coefficients. By a result due to B. Gustafsson [4], we may assume that the degree of P is equal to $2p$. Therefore, if $p = 2$, no straight line can intersect $E \cup \partial D$ in more than four points.

If $b \geqq 3$, we can find z_1 and z_2 inside of the inner boundary of D and a straight line passing through z_1 and z_2 intersects ∂D in at least six points. This is a contradiction. If $b = 2$ and $c \geqq 1$, then a straight line passing through z_1 inside of the inner boundary of D and a cusp intersects ∂D in at least five points by varying z_1 if necessary, a contradiction. If $b = 2$ and $s \geqq 1$, then a straight line passing through a point inside of the inner boundary of D and a singular point intersects $E \cup \partial D$ in at least five points, again a contradiction.

By Theorem, we get $s_{+1} - s_{-1} + c_{+1/2} = b$ if $p = 2$ and so $s_{+1} + c_{+1/2} \geqq b$. By the above argument, it follows that $b = 1$.

Next we shall show that if $p = 2$, then the index of a singular point is equal to $+1$ and that of a cusp is equal to $+1/2$. Let ς be a singular point of D. We may assume that $D = R(U)$ for a rational function R of order two which is univalent in the unit disk U, that $R(0) = 0$ and that $\varsigma = 0$. Since $S(z) = \overline{R}(1/w)$, ∞ is a zero of $\overline{R}$ and so 0 and ∞ are zeros of R. Hence we may assume that $R(w) = w/(1 + c_1 w + c_2 w^2)$, where $|c_2| < 1$. Since $w = R^{-1}(z) = z + 0(z^2)$ and $\overline{R}(1/w) = w/(w^2 + \overline{c}_1 w + \overline{c}_2) = (1/\overline{c}_2)w + 0(w^2)$, $S(z) = (1/\overline{c}_2)z + 0(z^2)$. Therefore $\varsigma = 0$ is a nondegenerate zero of v and the index is equal to $+1$, because $|1/\overline{c}_2| > 1$.

To show that every cusp has the index $+1/2$, we assume that $D = R(U)$ for a rational function R of order two which is univalent in U, that $R(0) = 0$, that $z = 0$ is one of the poles of S and that $R(-1)$ is a cusp on ∂D. Then 0 is a zero of R and ∞ is a pole of R and so we may assume that $R(w) = w(1 - b_2 w)/(1 - a_1 w)$, where $|a_1| < 1$, $0 < |b_2| < 1$ and $a_1 \neq b_2$. If $a_1 \neq 0$, then $R'(w) = a_1 b_2 (w^2 - (2/a_1)w + 1/(a_1 b_2))/(1 - a_1 w)^2$ and $R'(w) = 0$ has solutions -1 and $-1/(a_1 b_2)$. Hence $2/a_1 = -1 - 1/(a_1 b_2)$, namely,

$b_2 = -1/(2+a_1)$. If $a_1 = 0$, then $R(w) = w(1-b_2 w)$ and $R'(w) = 1-2b_2 w$. Since $R'(-1) = 0$, $b_2 = -1/2$. Therefore the equation $b_2 = -1/(2 + a_1)$ also holds in the case $a_1 = 0$. Let $w = w(t) = (t-1)/(t+1)$ be a one-to-one conformal mapping of $\{t;\ \mathrm{Re}\ t > 0\}$ onto U and set $F(t) = R(w(t))$. Then

$$F(t) = (1 + a_1)^{-1}(2 + a_1)^{-1}\{-(1 + a_1) + 4t^2 - 8(1 + a_1)^{-1}t^3 + 0(t^4)\}$$

and

$$\mathrm{Re}\{(1 + \bar{a}_1)(2 + \bar{a}_1)v(z)\} = 16r^3\{(\mathrm{Re}\frac{1}{1 + a_1}) \cos 3\theta + 0(r)\},$$

where $t = re^{i\theta}$, and so the cusp $R(-1)$ has the index $+1/2$, because $\mathrm{Re}\ 1/(1 + a_1) > 1/2 > 0$ if $|a_1| < 1$.

The above arguments show that $s_{-1} = s_0 = c_{-1/2} = 0$. Hence, by Theorem, $s_{+1} + c_{+1/2} = 1$ if $p = 2$.

Finally we shall show examples of quadrature domains of order two. Let $R(w) = w + a_2 w^2$. Then R is univalent in U if and only if $|a_2| \leq 1/2$. We take a_2 so that $|a_2| \leq 1/2$ and set $D = R(U)$. Then D is a quadrature domain of order two and $q = 1$. The case $c_{+1/2} = 1$ occurs if and only if $|a_2| = 1/2$.

To discuss the case $q = 2$, let $R(w) = w(1 - b_2 w)/(1 - a_1 w)$ and assume that $a_1 \neq 0$. Since $R(w_2) - R(w_1) = (w_2 - w_1)(1 - a_1 w_1)^{-1}(1 - a_1 w_2)^{-1}\{1 - b_2(w_1 + w_2) + a_1 b_2 w_1 w_2\}$, R is univalent in U and satisfies $R'(w) \neq 0$ on the closure of U if $0 < |a_1| < 1$ and $0 < |b_2| < 1/3$. The image $D = R(U)$ is a quadrature domain with $p = q = 2$ and there are no cusps on ∂D. Assume next that $0 < |a_1| < 1$ and $b_2 = -1/(2 + a_1)$. Since

$$|(1+w_1)+(1+w_2)|^2 - |1-w_1 w_2|^2 = |1+w_1|^2(1-|w_2|^2)+2\mathrm{Re}(1+w_1)|1+w_2|^2,$$

$|a_1(1 - w_1 w_2)| < |(1 + w_1) + (1 + w_2)| = |2 + w_1 + w_2|$ for a_1, w_1 and w_2 with $|a_1| < 1$, $|w_1| < 1$ and $|w_2| < 1$. Hence $1 - b_2(w_1 + w_2) + a_1 b_2 w_1 w_2 = 1 + (2 + a_1)^{-1}(w_1 + w_2) - a_1(2 + a_1)^{-1}w_1 w_2 = (2 + a_1)^{-1}\{(2 + w_1 + w_2) + a_1(1 - w_1 w_2)\} \neq 0$ for w_1 and w_2 in U and so $w_1 = w_2$ if $R(w_1) = R(w_2)$. Thus R is univalent in U and $R(-1)$ is a cusp of $D = R(U)$.

Summing up we see that if $p = 2$, then $s_{-1} = s_0 = c_{-1/2} = 0$ and $s_{+1} + c_{+1/2} = 1$. Two cases $c_{+1/2} = 0$ and $c_{+1/2} = 1$ occur in both cases $q = 1$ and $q = 2$.

Department of Mathematics; Tokyo Metropolitan University; Fukasawa, Setagaya-ku; Tokyo 158; Japan

REFERENCES

1. Aharonov, D. and Shapiro, H.S., *Domains on which analytic functions satisfy quadrature identities*, J. Anal. Math. **30** (1976), 39–73.
2. Brannan, D.A., *Coefficient regions for univalent polynomials of small degree*, Mathematika **14** (1967), 165–169.
3. Davis, P.J., "The Schwarz Function and its Applications," Carus Math. Monographs, No. 17, Math. Assoc. Amer., 1974.
4. Gustafsson, B., *Quadrature identities and the Schottky double*, Acta Appl. Math. **1** (1983), 209–240.
5. Sakai, M., "Quadrature Domains," Lecture Notes in Math. Vol. 934, Springer–Verlag, Berlin, 1982.
6. Suffridge, T.J., *Extreme points in a class of polynomials having univalent sequential limits*, Trans. Amer. Math. Soc. **163** (1972), 225–237.

The local modulus of continuity of
an analytic function

BY WAYNE SMITH AND DAVID A. STEGENGA

1. Introduction.

Let $f(z)$ be an analytic function on the unit disk D and let $\varsigma \in \partial D$. If $f(\varsigma)$ exists as a radial limit, then we define

$$\omega(f, t, \varsigma) = \sup\{|f(z) - f(\varsigma)| : |z - \varsigma| \le t, z \in D\}$$

and

$$\tilde{\omega}(f, t, \varsigma) = \text{ess sup } \{|f(e^{i\Theta}) - f(\varsigma)| : |e^{i\Theta} - \varsigma| \le t\}.$$

The usual modulus of continuity of f on D is $\omega(f, t) = \sup\{\omega(f, t, \varsigma) : \varsigma \in \partial D\}$ and the modulus of continuity of the restriction of f to ∂D is defined similarly as $\tilde{\omega}(f, t)$. See [RST].

It is known that there is a constant c such that $\omega(f, t) \le c\tilde{\omega}(f, t)$, whenever f is continuous on the closure of D and analytic on D. See [T] and [RST]. In [RST] it is also shown that the best possible value for c lies in the interval $(1, 3]$. A slightly more general way of expressing this theorem is by using an increasing function $\varphi(t)$ as a comparison. In other words, the hypothesis is that $\tilde{\omega}(f, t) \le \varphi(t)$ and the conclusion is that $\omega(f, t)$ is dominated by $c\varphi(t)$.

In [Se, p. 17] it is proved that for $\varphi(t) = t^\alpha$, with $0 < \alpha \le 1$, the constant c is one. In [RST], they give a proof using Hardy spaces. Their technique proves a more general result, namely, if $\tilde{\omega}(f, t, \varsigma) \le t^\alpha$ then $\omega(f, t, \varsigma) \le t^\alpha$, whenever f is bounded and analytic. Results of this type for functions f continuous on the closed disk and also more general domains are contained in [GHH], [H1], [RST] and [T]. All of these results contain restrictions on φ in addition to monotonicity. We give an example in a later section showing that this is necessary in order to obtain a bound for the ratio $\omega/\tilde{\omega}$. Our theorems are a generalization of this result to positive increasing functions φ, but the function f must satisfy some geometric restrictions. In addition, no continuity assumptions are needed. For the class of univalent functions we have the following theorem.

The second author is supported in part by a grant from the National Science Foundation.

THEOREM 1. *Let $\varphi(t)$ be a positive increasing function and $0 < t_0 < 2$. There is a constant $c_1(t_0) > 0$ such that, for all $\varsigma \in \partial D$ and all univalent functions f analytic in D for which $f(\varsigma)$ exists as a radial limit, if*

$$\tilde{\omega}(f, t, \varsigma) \leq \varphi(t) \quad \text{for } 0 \leq t \leq t_0,$$

then

$$\omega(f, t, \varsigma) \leq c_1(t_0)\varphi(t) \quad \text{for } 0 \leq t \leq t_0.$$

For $f(z)$ analytic on a domain G, let $d_G(f(z))$ denote the radius of the largest schlicht disk centered at the point $f(z)$ in the image Riemann surface of f over G. $\mathcal{B}_0$ is the space of functions analytic in D for which $d_D(f(z))$ tends to 0 as $|z|$ tends to 1. This is a subspace of the space of Bloch functions. See [**ACP**].

A family of functions $\mathcal{M}$ is said to be a linearly invariant family of bounded type if the following conditions hold:

(a) The functions in $\mathcal{M}$ are analytic and locally univalent in D and satisfy $f(0) = 0$ and $f'(0) = 1$;

(b) If $f \in \mathcal{M}$ and g is a Möbius transformation of D, then $[f(y(z)) - f(g(0))]/g'(0)f'(g(0)) \in \mathcal{M}$;

(c) There is a constant K so that for all $f \in \mathcal{M}$ $1/2\pi \int_0^{2\pi} \log^+ |f(re^{i\Theta})|d\Theta \leq K$, $0 \leq r < 1$.

The class of normalized univalent functions is a linearly invariant family of bounded type. Normalized locally univalent p-valent functions make up another such family [**P1**, Satz 1.3, p. 119]. Our next theorem concerns families of functions of this type.

THEOREM 2. *Let $\varphi(t)$ be a positive increasing function and $0 < t_0 < 2$. Suppose that $\mathcal{M}$ is a linearly invariant family of bounded type and that $\mathcal{M} \subset \mathcal{B}_0$. Then there is a constant $c_2(t_0) = c_2(t_0, \mathcal{M}) > 0$ such that, for all $\varsigma \in \partial D$ and all functions f with $[f(z) - f(0)]/f'(0) \in \mathcal{M}$ and with $f(\varsigma)$ existing as a radial limit, if*

$$\tilde{\omega}(f, t, \varsigma) \leq \varphi(t) \quad \text{for } 0 \leq t \leq t_0,$$

then

$$\omega(f, t, \varsigma) \leq c_2(t_0)\varphi(t) \quad \text{for } 0 \leq t \leq t_0.$$

The restriction that $t_0 < 2$ in Theorems 1 and 2 is necessary. An example will be presented in a later section to show this. The construction of this example can be motivated by two recent papers of A. Hinkkanen, [**H2**] and [**H3**].

Assume that φ is a positive increasing function in C^2, the function f is analytic in D and continuous in $\overline{D}$, and that $\tilde{\omega}(f, t, \varsigma) \leq \varphi(t)$ for $0 \leq t \leq 2$, where $\varsigma \in \partial D$. Let $M(t) = \log \varphi(e^t)$ for t real. Then by [**H2**, (2.6), p. 9] we have for $|z_0| < 1$ that

$$
\begin{aligned}
\log \frac{|f(z_0) - f(\varsigma)|}{\varphi(|z_0 - \varsigma|)} &\leq - \int_{r_0}^{2} M'(\log r)\, dG(r) \\
&= - \int_{r_0}^{2} \frac{dG(r)}{d \log r} dM(\log r),
\end{aligned}
$$

(1)

where $r_0 = |z_0 - \varsigma|$ and

$$
G(r) = 1/2\pi \int_0^{2\pi} g(\varsigma + re^{i\Theta}, z_0)\, d\Theta.
$$

Here $g(z, z_0)$ is the Green's function of the unit disk with pole at the point z_0 and $g(z, z_0) = 0$ for $|z| \geq 1$. Equation (1) is based on [**H3**, Lemma 1]. In [**H2**] and [**H3**] it is assumed that $\varphi(2t) \leq 2\varphi(t)$, but for (1) to hold we just need φ to be nice enough for all integrals involved to converge.

It is also shown in [**H2**, p. 9] that, for $r > r_0$, $G(r)$ is decreasing and is a convex function of $\log r$. We can now see from (1) that in order to make $|f(z_0) - f(\varsigma)|/\varphi(|z_0 - \varsigma|)$ large, we should let $\varphi(t)$ have a big jump quite soon after $t = r_0$. Furthermore, examples in [**H2**] (in particular the construction of the function g_1 in the proof of [**H2**, Lemma 2]) suggest that for (1) to be reasonably sharp, $|f(z) - f(\varsigma)|$ should be roughly equal to $\varphi(|z - \varsigma|)$ for $|z| = 1$. Thus the boundary values of f should take a big jump after $|z - \varsigma| = r_0$. We also have, as shown in [**H2**], that if φ satisfies a condition such as $\varphi(2t) \leq 2\varphi(t)$, then the right side of (1) is less than an absolute constant. Hence we should not have φ satisfying any condition of this type.

The above discussion indicates that to construct an example showing that $\omega(f, r_0, \varsigma)/\tilde{\omega}(f, r_0, \varsigma)$ can be arbitrarily large, one should arrange for $\tilde{\omega}(f, r_0 + \varepsilon, \varsigma)/\tilde{\omega}(f, r_0, \varsigma)$ to be larger than some arbitrarily chosen constant. If no restrictions are placed on f, then the required jump in the boundary values of f could be anywhere. However if geometric restrictions are placed on f as in Theorems 1 and 2, then these theorems show that any large jump must occur near the point antipodal to ς, so r_0 should be close to

2. A function with these properties is easily constructed by means of a conformal mapping onto a suitable simply connected domain.

We shall prove Theorems 1 and 2 in Section 2. In Section 3 the example showing the necessity of the restriction $t_0 < 2$ will be given. Finally, in Section 4 we present an example showing that if no restriction is placed on the function φ, then the function f must satisfy some geometric restrictions for the conclusion of the theorems to hold.

2. Proofs of the Theorems.

We will require two lemmas concerning linearly invariant families of bounded type.

LEMMA 1 (POMMERENKE, [**P1**, SATZ 2.8, P. 137]). *Let M be a linearly invariant family of bounded type and let $0 < \beta < 2\pi$. There is a constant $k_1(\beta) = k_1(\beta, M)$ such that if $f \in M$ and f is analytic in a neighborhood of $\overline{D}$, then*

$$\operatorname{diam} \{f(x) : 0 \le x \le 1\} \le k_1(\beta) \operatorname{diam} \{f(e^{i\Theta}) : 0 \le \Theta \le \beta\}.$$

LEMMA 2 (POMMERENKE, [**P1**, SATZ 2.11, P. 145]). *Let M be a linearly invariant family of bounded type and let $0 < \lambda \le 2\pi$. There is a constant $k_2(\lambda) = k_2(\lambda, M)$ such that if $f \in M$, $z_0 \in D$, $A \subset \partial D$ and the harmonic measure of A at z_0 is at least λ, then there is a non-euclidean segment S from z_0 to A with*

$$\operatorname{diam} f(S) \le k_2(\lambda) d_D(f(z_0)).$$

The proofs of the two theorems are nearly the same. We shall prove Theorem 1 and then just indicate the changes required to prove Theorem 2.

Let f be a univalent function. We assume that $\varsigma = 1$, that $f(1)$ exists as a radial limit, $f(1) = 0$ and that $0 < t_0 < 2$. Suppose that z is in D, has a nonnegative imaginary part and satisfies $|z - 1| = t \le t_0$. We assume that $\tilde{\omega}(f, t, \varsigma)$ is finite, otherwise the theorem is trivially true. Let C be the arc $\{e^{i\Theta} : 0 \le \Theta < \pi, |e^{i\Theta} - 1| \le t\}$. Then it is elementary that the harmonic measure of C with respect to the point z is greater than a positive constant $\beta_0/2\pi$ which depends only on t_0. We will assume that $f(\xi)$ exists as a nontangential limit, where C has endpoints 1 and ξ. Since this is true for almost all ξ, it is sufficient to consider this case.

There is a Möbius transformation g of D which fixes the point 1 and moves z to the origin. Since harmonic measure is preserved the transformation $\tilde{C}$ of the arc C will be the arc from 1 to $e^{i\beta}$, where $\beta \ge \beta_0$. Let

G be the sector bounded by the segment $[0,1]$, the circular arc $\tilde{C}$, and the radial segment $R = [0, e^{i\beta}]$. Let $\tilde{f}$ denote the transformation $f \circ g^{-1}$ of f. Since $\tilde{f}$ is univalent and its boundary values on $\tilde{C}$ are essentially bounded by $\tilde{\omega}(f, t, \varsigma)$, it is well known that $\tilde{f}$ is bounded on G.

Since $\tilde{f}(G)$ is bounded it follows that $d_G(\tilde{f}(z')) < \varphi(t)$, whenever $z' \in G$ and $|z'|$ is sufficiently close to one. For $0 < r < 1$, let $\tilde{C}_r = \{z \in G : |z| = r\}$. We choose r close enough to one so that $d_G(\tilde{f}(z')) < \varphi(t)$, whenever z' is in $\tilde{C}_r$. In addition, we assume that diam $\{\tilde{f}(se^{i\Theta}) : r \le s \le 1\} < \varphi(t)$ for $\Theta = 0, \beta$.

It is elementary that $\tilde{C}_r$ has a larger harmonic measure at the origin than $\tilde{C}$. By applying Lemma 1 to the function $[\tilde{f}(rz) - \tilde{f}(0)]/r\tilde{f}'(0)$, with $\mathcal{M}$ the class of normalized univalent functions, we obtain that $|\tilde{f}(0) - \tilde{f}(r)| \le k_1(\beta)$ diam $\tilde{f}(\tilde{C}_r)$. Let E be the part ∂G where the modulus exceeds r. It is fairly obvious that the harmonic measure $\omega(a, E, G)$ is bounded below by a positive absolute constant c_3, for any $a \in \tilde{C}_r$. Let a be a point on $\tilde{C}_r$. By applying Lemma 2 to the univalent function $\tilde{f}$ defined on G, we can find b in the set E, with $|\tilde{f}(a) - \tilde{f}(b)| \le c_4 d_G(\tilde{f}(a)) \le c_4\varphi(t)$, where c_4 is another absolute constant which we assume to be greater than one. In addition, if b is in $\tilde{C}$, then we may assume (using Lemma 2 again) that our hypothesis holds, i.e., that $|\tilde{f}(b)| \le \varphi(t)$. Thus, we have that the diameter of $\tilde{f}(\tilde{C}_r)$ is dominated by $5c_4\varphi(t)$. Finally, $|f(z) - f(\varsigma)|$ equals $|\tilde{f}(0)| \le (5k_1(\beta)c_4 + 1)\varphi(t) = c\varphi(|z - 1|)$ and the proof of Theorem 1 is complete.

The proof of Theorem 2 is essentially the same as that of Theorem 1. Now, of course, Lemma 1 and Lemma 2 are applied to $\mathcal{M}$, rather than to the class of normalized univalent functions. Also notice that if $f \in \mathcal{M}$, then f has bounded Nevanlinna characteristic, and so f has nontangential limits almost everywhere. In proving Theorem 2 it isn't necessary to show that $\tilde{f}$ is bounded on G. The inequality $d_G(\tilde{f}(z')) \le d_D(\tilde{f}(z')) < \varphi(t)$, whenever $z' \in G$ and $|z'|$ is sufficiently close to 1, is an immediate consequence of the hypothesis that $f \in \mathcal{B}_0$. Consequently the proof goes through as before.

REMARKS:

(a) An alternative proof of Theorem 1 could be based on Corollary 11.4 in [**P**, p. 344].

(b) If f is bounded and univalent, then there is a constant depending on f such that $\omega(f, t, \varsigma) \le c\tilde{\omega}(f, t, \varsigma)$, for all $t \le 2$. One may take c to be the maximum of $\tilde{\omega}(f, 2, \varsigma)/\tilde{\omega}(f, 1, \varsigma)$ and $c_1(t_0)$, where $t_0 = 1$.

138

3. Example One.

The restriction that t_0 be strictly less than two in the theorems is essential. An example will be constructed below. We will require some ideas concerning the hyperbolic metric and noneuclidean geometry for its proof.

Denote by $\rho_D(z_1, z_2)$ the hyperbolic (or noneuclidean) distance between z_1 and z_2 in D defined by (see [**A**], p. 2)

$$\rho_D(z_1, z_2) = \inf\left\{\int_\Gamma \frac{2|dz|}{1 - |z|^2} : \Gamma \text{ an arc from } z_1 \text{ to } z_2 \text{ in } D\right\}$$
$$= \log\frac{|1 - \bar{z}_1 z_2| + |z_1 - z_2|}{|1 - \bar{z}_1 z_2| - |z_1 - z_2|}.$$

This distance is invariant under conformal self-mappings of the disk. The geodesics in this metric are the noneuclidean segments in D, in other words, the family of circles orthogonal to the unit circle. The corresponding distance on G is $\rho_G(w_1, w_2) = \rho_D(z_1, z_2)$ where $w_i = f(z_i)$ for $i = 1, 2$, where G is a simply connected domain and f is any conformal mapping of D onto G. There is a function $h(w)$ on G satisfying

$$h(w)|dw| = \frac{2|dz|}{1 - |z|^2}, \quad w = f(z),$$

for all mapping functions f. Thus, ρ_G can be computed by integrating $h(w)$ over arcs in G. Using the notation $\delta(w)$ for the Euclidean distance to the boundary of G, we have the basic inequality $\delta(f(z)) \leq |f'(z)|(1 - |z|^2) \leq 4\delta(f(z))$ for z in D ([**P**, p. 22]) and hence $1/2 \leq h(w)\delta(w) \leq 2$ for w in G. This allows us to estimate the hyperbolic distance in G, by using the geometry of G.

We are now ready to construct our first example. Let $a > 1$ and $0 < b < 1$ be given constants. Define a Jordan region G by the union

$$G = [-1, 1] \times [-1, 1] \cup [1, 2] \times [-b, b] \cup [2, 2a + 2] \times [-a, a].$$

Let f be the Riemann mapping function which maps D onto G, with $f(0) = 0$ and $f(-1) = -1$. By a symmetry argument we see that $f(e^{i\alpha}) = 2 + ib$ and $f(e^{-i\alpha}) = 2 - ib$ for some $\alpha > 0$. Since a square has equal harmonic measure on each of its sides (with respect to its center) we see that $\alpha < \pi/4$.

Let C_1 be the noneuclidean segment in D connecting $e^{i\alpha}$ to $e^{-i\alpha}$. Then C_1 intersects the real axis at a point r_1, where $0 < r_1 < 1$. By a theorem of Gehring and Hayman [**GH**, Theorem 2, p. 357], there is an absolute constant c_1 such that the length of $f(C_1)$, denoted by $|f(C_1)|$, is less than $c_1 b$. We will assume that $c_1 b \leq 1/2$. If $x_1 = f(r_1)$, then it follows that $x_1 \in [2 - c_1 b, 2 + c_1 b]$. We obtain a lower bound for $\rho_D(0, r_1)$ as follows:

$$\rho_D(0, r_1) = \rho_G(0, x_1) \geq \int_1^{2-c_1 b} h_G(x)\, dx$$

$$\geq \frac{1}{2} \int_1^{2-c_1 b} \frac{dx}{b} = \frac{1 - c_1 b}{2b}$$

$$\geq \frac{1}{4b} \, .$$

Let C_2 be the smaller arc connecting $e^{i\alpha}$ and $e^{-i\alpha}$ on the circle centered at $z = -1$. Let r_2 denote the point where C_2 intersects the real axis. Clearly, $r_1 < r_2 < 1$. An elementary argument shows that there is an $\varepsilon > 0$, which is independent of a and b, such that

$$\rho_D(r_1, r_2) \geq \varepsilon \rho_D(0, r_1).$$

Let $r = f^{-1}(a + 2)$. Since $f(r_1) \leq 2 + c_1 b \leq 3$, we obtain that $r_1 \leq r$. If we show that $\rho_D(r_1, r) \leq \varepsilon \rho_D(0, r_1)$, then the above relation would imply that $r \leq r_2$.

We compute an upper bound for $\rho_D(r_1, r)$ as follows:

$$\rho_D(r_1, r) \leq \rho_G(x_1, 2 + c_1 b) + \rho_G(2 + c_1 b, a + 2)$$

$$\leq 4c_1 + 2 \int_{c_1 b}^a \frac{dt}{t}$$

$$\leq 3 \log \frac{a}{b} \, ,$$

where the last inequality requires that b be sufficiently small. Combining our estimates we see that the inequalities $r_1 \leq r \leq r_2$ hold whenever b is sufficiently small and $\log a \leq (\varepsilon/12) b^{-1}$. Clearly, this condition can be achieved for any value of a by choosing b small enough.

Finally, suppose that a is given and b is chosen so that $r_1 \leq r \leq r_2$ in the above construction. Let $\varsigma = -1$. Then, $\tilde{\omega}(f, 1 + r_2, \varsigma) \leq 4$ and yet $\omega(f, 1 + r_2, \varsigma) \geq a + 3$. This proves our assertion that the theorems would be false without the hypothesis that $t_0 < 2$.

4. Example Two.

If the function φ is not assumed to have regular growth, then the geometric restrictions on the function f are necessary in Theorem 1 and in

Theorem 2. Without these restrictions the theorems would not hold, even for continuous functions. To show this we will first construct a function g which is continuous for $y \geq 0$, analytic for $y > 0$ and for which

$$
(2) \qquad \limsup_{y \to 0+} \frac{|g(0) - g(iy)|}{\tilde{\omega}(g, y, 0)} = \infty .
$$

We start by constructing a function ψ on $(-\infty, \infty)$. Let $a > 4$ be a large number. If $n = 0, 1, \ldots$ and $a^{-n-1} < |t| \leq a^{-n}$, then let $\psi(t) = 2^n$. Let $\psi(t) = 0$ otherwise. In this example the role of the function $M(t)$ in equation (1) will be taken by $-\psi(e^t)$. Thus equation (1) indicates why a function such as $\psi(t)$ with arbitrarily large jumps should be considered.

A straightforward estimate shows that ψ is integrable on $(-\infty, \infty)$ and that for $n = 0, 1, \ldots$ the inequality

$$
(3) \qquad \frac{2}{\pi} \int_0^{a^{-n-1}} \frac{a^{-n}}{x^2 + a^{-2n}} \psi(x)\,dx \leq \frac{2a^n}{\pi} \int_0^{a^{-n-1}} \psi\,dx \leq \frac{8}{\pi} \frac{2^n}{a}
$$

holds. On the other hand, explicit integration shows that

$$
\frac{2}{\pi} \int_{a^{-n-1}}^{\infty} \frac{a^{-n}}{x^2 + a^{-2n}} \psi(x)\,dx
$$

$$
\leq \frac{2}{\pi} \left(\int_{a^{-n-1}}^{a^{-n}} \frac{2^n a^{-n}}{x^2 + a^{-2n}}\,dx + \int_{a^{-n}}^{\infty} \frac{2^{n-1} a^{-n}}{x^2 + a^{-2n}}\,dx \right)
$$

$$
\leq \frac{2}{\pi} 2^n \left(\arctan 1 + \frac{1}{2}\left(\frac{\pi}{2} - \arctan 1\right) \right) = \frac{3}{4} 2^n
$$

and hence if a is sufficiently large, the inequality (3) and the above imply that

$$
(4) \qquad \lim_{n \to \infty} \left(\psi(a^{-n}) - \frac{1}{\pi} \int_{-\infty}^{\infty} \frac{a^{-n}}{x^2 + a^{-2n}} \psi(x)\,dx \right) = \infty .
$$

Furthermore, it is clear that ψ can be modified so that it is continuously differentiable everywhere except at zero and satisfies (4). We assume that ψ has been modified in this manner.

Now we let $g(z)$ be the outer function with

$$
(5) \qquad \log |g(z)| = -\frac{1}{\pi} \int_{-\infty}^{\infty} \frac{y}{(x-t)^2 + y^2} \psi(t)\,dt
$$

where $z = x + iy$ and $y > 0$. It is well known that g is continuous for $y \geq 0$, that $g(0) = 0$ and that $|g(x)| = \exp(-\psi(x))$ for $-\infty < x < \infty$. Now (2) is an immediate consequence of (4) and (5). This completes the construction for the upper half plane. Clearly, this example can be transferred to the disk.

Department of Mathematics, University of Hawaii, 2565 The Mall, Honolulu, Hawaii 96822

References

[A] Ahlfors, L.V., "Conformal Invariants: Topics in Geometric Function Theory," McGraw–Hill Book Company, 1973.

[ACP] Anderson, J.M., Clunie, J. and Pommerenke, Ch., *On Bloch functions and normal functions*, J. reine angew. Math. **270** (1974), 12–37.

[GH] Gehring, F.W. and Hayman, W.K., *An inequality in the theory of conformal mapping*, J. Math. Pures Appl. (9) **41** (1962), 353–361.

[GHH] Gehring, F.W., Hayman, W.K. and Hinkkanen, A., *Analytic Functions Satisfying Hölder Conditions on the Boundary*, J. Approx. Theory **35** (1982), 243–249.

[H1] Hinkkanen, A., *On the Modulus of Continuity of Analytic Functions*, Ann. Acad. Sci. Fenn., Ser. AI Math. **10** (1985), 247–253.

[H2] ______________, *The sharp form of certain majorization theorems for analytic functions*, MSRI preprint 10719-86, 1–34.

[H3] ______________, *On the majorization of analytic functions*, to appear in Indiana Univ. Math. J.

[P] Pommerenke, Ch., "Univalent Functions," Vanderhoeck & Ruprecht in Gottingen, 1975.

[P1] ______________, *Linear-invariante Familien analytischer Funktionen I*, Math. Ann. **155** (1964), 108–154.

[RST] Rubel, L.A., Shields, A.L. and Taylor, B.A., *Mergelyan Sets and the Modulus of Continuity of Analytic Functions*, J. Approx. Theory **15** (1975), 23–40.

[Se] Sewell, W.E., "Degree of Approximation by Polynomials in the Complex Domain," Princeton University Press, Princeton, 1942.

[T] Tamrazov, P.M., *Contour and solid structure properties of holomorphic functions of a complex variable*, Russ. Math. Surveys **28** (1973), 141–173.

Quasiconformal isotopies

BY CLIFFORD J. EARLE AND CURT MCMULLEN

INTRODUCTION

Let X be a hyperbolic Riemann surface or orbifold, possibly of infinite topological complexity. Let $\phi : X \to X$ be a quasiconformal map. We show the following conditions are equivalent (§1):

(a) ϕ has a lift to the universal cover Δ which is the identity on $\mathbf{S}^1$;

(b) ϕ is homotopic to the identity rel the ideal boundary of X; and

(c) ϕ is isotopic to the identity rel ideal boundary, through uniformly quasiconformal maps.

(A related result in the PL category was established by Epstein [**Eps**].) The proof relies on the barycentric extension introduced by Douady and Earle [**DE**]. Applications include the equivalence of alternative definitions of Teichmüller space and new proofs of the Bers-Greenberg theorem and of the contractibility of $Diff_0(X)$.

To understand the relation of condition (b) to other notions of relative homotopy, we study the geometry of the universal covering map $\pi : \Delta \to X$. For a subdomain of the Riemann sphere, we show a homotopy rel the frontier of X is a homotopy rel ideal boundary (but the converse is false for certain non-locally connected domains) (§2). We show a homotopy rel ideal boundary lifts to a homotopy rel the sphere at infinity (but this fails for hyperbolic 3-manifolds) (§3). Finally, under the assumption of uniform quasiconformality, all notions of relative isotopy coincide. The proofs use harmonic measure and hyperbolic geometry.

These results have applications to the deformation theory of rational maps and Kleinian groups.

§1. Riemann Surfaces and Orbifolds.

We will work in the category of hyperbolic orbifolds, in the interest of obtaining the Bers-Greenberg theorem.

DEFINITIONS: Let Γ be a Fuchsian group, that is a discrete subgroup of conformal automorphisms of the unit disk Δ; we do not require Γ to be torsion-free or finitely generated. Then the quotient $X = \Delta/\Gamma$ has the

---

This work was supported in part by NSF grant DMS-8601016 (first author).

structure of a hyperbolic orbifold; each point of X has a neighborhood which is modelled on the quotient of a disk by a finite group of rotations. The projection $\pi : \Delta \to \Delta/\Gamma = X$ is a covering map of orbifolds; it is the *universal covering* of the orbifold X. Let $B \subset X$ denote the discrete set of *branch points* of X, i.e. the points in the quotient corresponding to fixed points of elliptic elements of Γ. Then $X - B$ has the structure of an ordinary Riemann surface.

By definition, a *continuous map* $\phi : X \to X$ is a map which is covered by a continuous map on the universal cover of X. In other words there must exist a continuous map $\hat{\phi} : \Delta \to \Delta$ such that $\phi \circ \pi = \pi \circ \hat{\phi}$. The map $\hat{\phi}$ is unique up to composition on the right and the left with elements of Γ.

We say ϕ is *the identity on* $\mathbf{S}^1$ if some choice of $\hat{\phi}$ can be completed to a continuous map of the closed disk $\overline{\Delta}$ pointwise fixing its boundary $\mathbf{S}^1$.

We say ϕ is *quasiconformal* or *conformal*, if $\hat{\phi}$ is quasiconformal or conformal (this property is clearly independent of the choice of $\hat{\phi}$).

Let $\Omega \subset \mathbf{S}^1$ denote the complement of the limit set of Γ; then the quotient $(\Delta \cup \Omega)/\Gamma$ is an orbifold with interior X and boundary Ω/Γ, which we call the *ideal boundary* of X (denoted ideal-∂X).

Let I denote the unit interval $[0,1]$, and consider a homotopy $\phi : I \times X \to X$ (which we denote by $\phi(t,x)$ or $\phi_t(x)$), such that $\phi_0 = id$. By definition, such a homotopy is covered by a homotopy on the universal cover of X.

Then ϕ_t is a homotopy *rel ideal boundary* if it can be completed to a homotopy of $(X \cup \text{ideal}-\partial X)$ pointwise fixing the ideal boundary. Similarly ϕ_t is a homotopy *rel* $\mathbf{S}^1$ if it has a lift $\hat{\phi}_t$ to a homotopy of the universal cover Δ of X which can be completed to a homotopy of $\overline{\Delta}$ pointwise fixing $\mathbf{S}^1$.

REMARK: It is clear that a homotopy rel $\mathbf{S}^1$ is a homotopy rel ideal boundary; the converse will be established in §3.

THEOREM 1.1. *Let* $\phi : X \to X$ *be a* K-*quasiconformal map. Then the following conditions are equivalent:*

(a) ϕ *is the identity on* $\mathbf{S}^1$.

(b) ϕ *is homotopic to the identity rel the ideal boundary of* X.

(c) ϕ *is isotopic to the identity rel ideal boundary, through* K'-*quasiconformal maps (where* K' *depends only on* K*).*

REMARKS:

(1) We shall see in §§2 and 3 that for a uniformly quasiconformal isotopy, the conditions (i) bounded, (ii) rel $\mathbf{S}^1$, (iii) rel ideal boundary and (iv) rel frontier (for a subdomain of $\hat{\mathbf{C}}$) are all equivalent. Thus the isotopy of (c) enjoys all these properties.

(2) The isotopy that we will construct in the proof respects the symmetries of ϕ. More precisely, let G be a group of conformal automorphisms of X, or more generally a semigroup of self-coverings of X. Then if $\phi \circ g = g \circ \phi$ for all g in G, the same will be true for each map ϕ_t occuring in the isotopy. This is useful for applications to rational maps and Kleinian groups.

PROOF OF THEOREM 1.1: The implications $(c) \Rightarrow (b) \Rightarrow (a)$ are easy to check.

To see $(a) \Rightarrow (c)$, let Γ be a Fuchsian group uniformizing X, and let $\phi : X \to X$ be a quasiconformal map which is the identity on $\mathbf{S}^1$. Then ϕ has a lift $\psi : \Delta \to \Delta$ that extends continuously to the identity on $\mathbf{S}^1$ and commutes with every element of Γ.

Let μ denote the dilatation of ψ, and let α_t denote the unique quasiconformal map of the disk to itself with dilatation $t\mu$, fixing $(1, i, -1)$. By Ahlfors-Bers [**AB**], α_t gives an isotopy of $\overline{\Delta}$, but *not* necessarily rel $\mathbf{S}^1$. Since μ is Γ-invariant,

$$\Gamma_t = \alpha_t^{-1} \circ \Gamma \circ \alpha_t$$

is a family of Fuchsian groups isomorphic to Γ. Also $\alpha_0 = \alpha_1 = id$ on $\mathbf{S}^1$, so $\Gamma_0 = \Gamma_1 = \Gamma$.

Let $\beta_t = ex(\alpha_t^{-1})$ denote the barycentric extension of the boundary values of α_t^{-1} (see Douady-Earle [**DE**]). By [**DE**] these extensions are K'-quasiconformal for K' depending only on K; they depend continuously on t, so β_t gives an isotopy of the closed disk; and by conformal naturality they conjugate the action of Γ_t to Γ throughout the entire unit disk. The initial and terminal maps β_0 and β_1 are the identity since they are barycentric extensions of the identity.

Let $\psi_t = \beta_t \circ \alpha_t$. Then ψ_t is Γ-equivariant, so it descends to an isotopy $\phi_t : X \to X$ connecting the identity map to ϕ. By construction β_t and α_t^{-1} agree on $\mathbf{S}^1$, so ψ_t is an isotopy rel $\mathbf{S}^1$ (and hence rel ideal boundary).

The argument shows ψ_t is in fact compatible with the group of all Möbius transformations commuting with ψ, justifying remark (2) above. $\square$

Applications and Refinements.

(1) Recall Bers' construction of the Teichmüller space $\mathrm{Teich}(X)$ [**B1**]: $\mathrm{Teich}(X)$ consists of pairs (Y, α) such that $\alpha : X \to Y$ is a quasiconformal homeomorphism of orbifolds, modulo the equivalence relation $(Y, \alpha) \sim (Z, \beta)$ if there is a *conformal* map $\gamma : Y \to Z$ such that

$$\phi = \beta^{-1}{\circ}\gamma{\circ}\alpha : X \to X$$

is the identity on $\mathbf{S}^1$.

By the theorem, if we replace this condition by the requirement: *ϕ admits a uniformly quasiconformal isotopy to the identity rel ideal boundary*, we obtain the same equivalence relation. Thus we have an equivalent definition of $\mathrm{Teich}(X)$ which is somewhat more intrinsic to X.

(Actually Bers does not define the Teichmüller space of X when X has branch points; however $\mathrm{Teich}(X)$ as defined above is always isomorphic to a space he does define, namely $\mathrm{Teich}(\Gamma)$, where Γ is a Fuchsian group uniformizing the orbifold X.)

(2) Using this remark, we have an independent demonstration of the Bers-Greenberg theorem [**BG**] (compare Marden [**Mar**] and Gardiner [**Gar**]):

COROLLARY 1.2 (BERS-GREENBERG). *Let X be a hyperbolic orbifold with branch set B. Then the Teichmüller spaces $\mathrm{Teich}(X)$ and $\mathrm{Teich}(X - B)$ are canonically isomorphic.*

PROOF: Restriction from X to $X - B$ defines a canonical bijection from quasiconformal homeomorphisms with domain X to those with domain $X - B$.

A uniformly quasiconformal isotopy to the identity on $X - B$ can be completed to an isotopy of the underlying topological space of X, which fixes B pointwise and is therefore an isotopy with respect to the orbifold structure. Conversely an isotopy to the identity on X fixes B pointwise and hence restricts to an isotopy on $X - B$.

The bordered Riemann surface $(X - B) \cup \mathrm{ideal}{-}\partial(X - B)$ is canonically identified with $(X \cup \mathrm{ideal} - \partial X) - B$. This can be demonstrated by factoring the universal covering of the hyperbolic Riemann surface $X - B$ through the universal covering $\Delta \to X$. Thus one of the isotopies is rel ideal boundary iff the other one is.

Using the isotopy definition of Teichmüller space described in point (1) above, we see two maps α and β are equivalent in $\mathrm{Teich}(X)$ if and only if their restrictions to $X - B$ are equivalent in Teich$(X - B)$, so we have established the desired bijection. $\square$

(3) Let X be a compact Riemann surface of genus $g \geq 2$ and $\Delta \to X = \Delta/\Gamma$ a universal covering. The ideal boundary of X is empty, and every diffeomorphism $\phi : X \to X$ is quasiconformal. By a theorem of Earle and Eells [**EE2**], the group $Diff_0(X)$ of diffeomorphisms $\phi : X \to X$ homotopic to the identity (with its C^∞ topology) is contractible. By Theorem 1.1, $Diff_0(X)$ consists of the diffeomorphisms ϕ that are the identity on $\mathbf{S}^1$. The isotopy ϕ_t constructed above depends continuously on both t and ϕ and provides an explicit contraction of $Diff_0(X)$ to the identity map.

(4) In general the complex dilatation of the isotopy ϕ_t constructed above need not vary continuously (in L^∞) as a function of t. Here is a more indirect proof of the implication $(a) \Rightarrow (c)$, which yields a stronger result:

THEOREM 1.3. *The isotopy in part (c) of Theorem 1.1 can be chosen so the complex dilatation μ_t of ϕ_t varies continuously in $M(X)$, the unit ball in the Banach space of measurable Beltrami differentials on X with the L^∞ norm.*

PROOF: In [**DE**] the barycentric extension is used to prove that the Teichmüller space of any Riemann surface or orbifold is contractible. By Earle and Eells [**EE1**], the map from $M(X)$ to Teich(X) obtained by solving the Beltrami equation is a locally trivial fibration, and the contractibility of Teich(X) implies the contractibility of the fiber F lying over (X, id).

By definition, a quasiconformal map $\phi : X \to X$ is the identity on $\mathbf{S}^1$ if and only if the complex dilatation $\mu(\phi)$ lies in F. Let μ_t be a path in F connecting 0 (the complex dilatation of the identity map) to μ. Then the corresponding quasiconformal isotopy ϕ_t satisfies the conditions of (c), and the complex dilatation varies continuously by construction. $\square$

REMARK: Since the complex dilatation of ϕ_t varies continuously, the equation

$$\phi = \phi_1 = \phi_{1/n} \circ [\phi_{1/n}^{-1} \circ \phi_{2/n}] \circ \ldots \circ [\phi_{n-1/n}^{-1} \circ \phi_1]$$

exhibits a factorization of ϕ into quasiconformal maps of small dilatation (for n sufficiently large), each of which is the identity on $\mathbf{S}^1$.

§2. Planar Domains.

We now specialize to the case where X is a subdomain of the Riemann sphere. To avoid confusion with the ideal boundary, we will refer to the topological boundary of $X \subset \hat{\mathbf{C}}$ as the *frontier* of X (denoted ∂X). We say $\phi_t : X \to X$ is a homotopy rel frontier if it can be completed to a homotopy of the closure $\overline{X}$ fixing ∂X pointwise.

Let $\pi : \Delta \to X$ denote the universal covering map. Our discussion of isotopies depends on a sort of uniform continuity for the inverse of π, which may be of interest in its own right.

Let $\gamma : I \to X$ be a path, $\hat{\gamma}$ a lift of γ to a path in Δ. Let diam(γ) denote the diameter of $\gamma(I)$ in the spherical metric, and diam($\hat{\gamma}$) the diameter of the lift in the Euclidean metric on the disk.

LEMMA 2.1. *There is a function* $\alpha(t) \to 0$ *as* $t \to 0$ *such that*

$$\mathrm{diam}(\hat{\gamma}) \leq \alpha(\mathrm{diam}(\gamma))$$

for all paths γ *in* X *and all choices of lifts* $\hat{\gamma}$.

REMARK: The lemma is really a comparison between metrics, which can be rephrased as follows. Let $p : \hat{X} \to X$ denote a topological universal covering for X. Define the *path metric* $d(x,y)$ on $\hat{X}$ by

$$d(x,y) \;=\; \inf\{\mathrm{diam}\, p(\gamma) : \gamma \text{ is a path joining } x \text{ to } y \text{ in } \hat{X}\}.$$

A choice of basepoints determines a homeomorphism $\pi^{-1}{\circ}p : \hat{X} \to \Delta$. The lemma states that this homeomorphism is uniformly continuous from the path metric to the Euclidean metric on the disk.

Using the Lemma above, we will prove

THEOREM 2.2. *A homotopy to the identity rel frontier is also a homotopy rel ideal boundary.*

REMARKS:

(1) This result is an important technical point in Sullivan's proof of the no wandering domains theorem for rational maps [**Sul**] and a parallel proof of Ahlfors' finiteness theorem [**B3**] (see also [**BR**], Lemma II of §5). In those works less general versions of the above are established by quite different arguments.

(2) The converse of Theorem 2.2 is false. For example, consider a domain with a comb in the boundary, such as X = the upper half-plane with the vertical segments of unit length lying above $z = 1, 1/2, 1/3, \ldots, 0$ removed. Define an isotopy on each rectangle $[1/(n+1), 1/n] \times [0,1]$ starting at the

identity, fixing the boundary throughout, and moving some interior point a definite vertical distance (say $1/2$). Extend by the identity to an isotopy of the whole region X. The result is an isotopy rel ideal boundary (since the diameters of pre-images of the rectangles in the universal cover Δ tend to zero in the Euclidean metric), but not rel ∂X (no continuous extension is possible near the vertical segment over $z = 0$). (See Figure 2.1.)

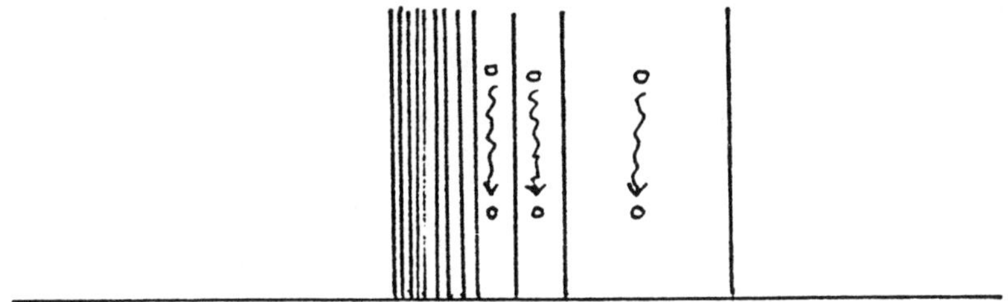

Figure 2.1. An isotopy rel ideal boundary but not rel ∂X.

This example cannot be made quasiconformal. We say an isotopy is *bounded* if it moves points only a uniformly bounded distance in the Poincaré metric; more precisely, each path $\phi(I, x)$ has a lift to Δ whose Poincaré diameter is bounded independent of x.

PROPOSITION 2.3.

(a) *A K-quasiconformal isotopy rel $\mathbf{S}^1$ is a bounded isotopy.*
(b) *A bounded isotopy is an isotopy rel frontier.*

PROOF: (a) A K-quasiconformal map fixing the boundary of the unit disk moves points in the interior only a uniformly bounded distance in the Poincaré metric. (b) Since the ratio of Poincaré metric to spherical metric tends to infinity as one nears the frontier of X, the isotopy extends continuously to the identity on ∂X. $\square$

COROLLARY 2.4. *Let $\phi : X \to X$ be K-quasiconformal. The following are equivalent.*

(a) *ϕ is the identity on $\mathbf{S}^1$.*
(b) *ϕ is homotopic to the identity rel the frontier of X.*

(c) There is a K'-quasiconformal isotopy ψ_t of the whole Riemann sphere, such that ψ_t fixes $\hat{\mathbf{C}} - X$ pointwise and its restriction to X provides an isotopy connecting ϕ to the identity.

PROOF: Clearly $(c) \Rightarrow (b)$. By Theorem 2.2 the homotopy of (b) is also rel ideal boundary, so $(b) \Rightarrow (a)$. Now assuming (a), Theorem 1.1 yields a K'-quasiconformal isotopy ϕ_t connecting ϕ to the identity. By Proposition 2.3 this extends to an isotopy rel frontier, which extends by the identity to an isotopy of the whole Riemann sphere. By a well-known result (see [**B2**] or [**DH**, Lemma 2]), the continuous extension of a quasiconformal map by the identity is still quasiconformal with the same dilatation, establishing (c). $\qquad\qquad\square$

We give the proof of 2.2 assuming the lemma on diameter of paths, to motivate the latter.

PROOF OF THEOREM 2.2: Let $\phi_t : \overline{X} \to \overline{X}$ be an isotopy connecting ϕ to the identity through maps fixing ∂X. Let $\hat{\phi}_t$ be the unique lift to an isotopy of Δ connecting a lift of ϕ to the identity. We claim $\hat{\phi}_t$ can be completed to an isotopy of $\overline{\Delta}$ fixing $\mathbf{S}^1$.

It suffices to construct for each $\epsilon > 0$ a neighborhood U of $\mathbf{S}^1$ such that $\text{diam}\hat{\phi}(I, z) < \epsilon$ for all z in U. To this end choose a neighborhood V of ∂X such that $\text{diam}\phi(I, x) < \delta$ for all x in V, where $\alpha(\delta) < \epsilon$ and α is the function provided by Lemma 2.1. Such a neighborhood exists because the isotopy fixes ∂X.

If $\pi(z)$ lies in V, then $\text{diam}(\hat{\phi}(I, z)) < \epsilon$ since it covers the path $\phi(I, x)$. On the other hand, $X - V$ is a compact set, so if $\pi(z)$ lies in $X - V$ the Poincaré diameter of $\hat{\phi}(I, z)$ is bounded above by a constant depending only on V. Since the Poincaré metric tends to infinity relative to the Euclidean metric as z tends to $\mathbf{S}^1$, there is a neighborhood U of $\mathbf{S}^1$ such that $\text{diam}\hat{\phi}(I, z) < \epsilon$ even if $\pi(z)$ lies in $X - V$.

It follows that $\hat{\phi}_t(z)$ tends to the identity uniformly as $|z| \to 1$, and so may be completed to an isotopy of $\overline{\Delta}$ fixing $\mathbf{S}^1$. Thus ϕ_t itself is an isotopy rel ideal boundary. $\qquad\qquad\square$

PROOF OF LEMMA 2.1: Let $\pi : \Delta \to X$ denote the universal covering map. We may assume $X \subset \hat{\mathbf{C}} - \{0, 1, \infty\}$ by applying a Möbius transformation (this changes spherical diameters by only a bounded factor.) Let $w = \pi(0)$ and let $\lambda : \Delta \to \hat{\mathbf{C}} - \{0, 1, \infty\}$ be the universal covering map normalized so that $\lambda(0) = w$. Then we may factor the map π as

$\pi = \lambda \circ \alpha$, where $\alpha : \Delta \to \Delta$ and $\alpha(0) = 0$.

Let γ be a path in X, $\hat{\gamma}$ a lift of γ to Δ. We wish to show that $\mathrm{diam}(\hat{\gamma})$ is controlled by $\mathrm{diam}(\gamma)$. This is clear when γ lies in a closed ball B centered at w and contained in X; in fact all branches of π^{-1} have uniformly bounded derivative throughout B, so in this case we have $\mathrm{diam}(\hat{\gamma}) = \mathbf{O}(\mathrm{diam}(\gamma))$.

Now assume γ is a definite distance from w. Let γ_0 be the lift $\alpha(\hat{\gamma})$ of γ to Δ via λ. For E a closed subset of the unit disk, let $\omega(z, E)$ denote the harmonic measure of E, i.e. the unique harmonic function on $\Delta - E$ with boundary values 1 on E and 0 elsewhere. Equivalently, $\omega(z, E)$ is the probability that a random path initiated at z hits E before exiting the unit disk.

To complete the proof, it suffices to establish the following chain of inequalities:

(i) $\mathrm{diam}(\hat{\gamma}) \leq \mathbf{O}(\omega(0, \hat{\gamma}))$

(ii) $\omega(0, \hat{\gamma}) \leq \omega(0, \gamma_0)$

(iii) $\omega(0, \gamma_0) \leq \mathbf{O}(1/\log(1/\mathrm{diam}(\gamma)))$

since together they imply that the size of $\hat{\gamma}$ is controlled by the size of γ.

To see that (i) is plausible, one can argue that the probability of hitting a set of given diameter is least when the set is near the boundary of the disk, and the inequality is clear for a subset of the boundary. For a precise argument and a sharper inequality, we refer to the paper of Fitzgerald, Rodin and Warschawski [**FRW**].

Inequality (ii) follows from standard monotonicity properties of harmonic measure; there are more paths leading to γ on the triply punctured sphere than on X, and any one in the correct homotopy class to hit $\hat{\gamma}$ is also in the correct homotopy class to hit γ_0. More formally, (ii) follows from the maximum principle, by comparing the boundary values of $\omega(z, \hat{\gamma})$ and $\omega(\alpha(z), \gamma_0)$ on $\Delta - \alpha^{-1}(\gamma_0)$.

Finally, inequality (iii) follows from a direct computation on the triply punctured sphere. The derivative of any branch of λ^{-1} (from the spherical to Euclidean metric) is $\mathbf{O}(1/d)$, where d is the distance to the nearest puncture. Thus, if γ is not contained in a ball of radius $\sqrt{\mathrm{diam}(\gamma)}$ about some puncture, the diameter of its lift γ_0 is $\mathbf{O}(\sqrt{\mathrm{diam}(\gamma)})$. Then (iii) follows from estimate $\omega(0, \gamma_0) = \mathbf{O}(1/\log(1/\mathrm{diam}(\gamma_0)))$, which holds for any path in the unit disk which is a definite distance from zero.

On the other hand, if γ is contained in the ball of radius $\sqrt{\mathrm{diam}(\gamma)}$ centered at a puncture, then its lift γ_0 is contained in a horoball of diameter

$O(1/\log(1/\mathrm{diam}(\gamma)))$, and (iii) follows from the fact that the harmonic measure of a horoball is comparable to its diameter. $\qquad\square$

REMARKS:

(1) It follows from the proof that we may take $\alpha(t) = O(1/\log(1/t)))$ (and this order of magnitude is sharp for the triply punctured sphere).

(2) If X is simply connected we may take $\alpha(t) = O(\sqrt{t})$ (and this is sharp for the complement of a slit). To improve the estimate in the simply connected case, one applies the Beurling projection theorem (see Ahlfors [Ahlf]) to replace (ii) and (iii) with the inequality $\omega(0,\gamma) \leq O(\sqrt{\mathrm{diam}(\gamma)})$. We are grateful to Peter Jones for a discussion of this point.

§3. Ideal Boundary.

We conclude with a parallel discussion in which the frontier of X is replaced by its ideal boundary.

Let $\overline{X} = X \cup \mathrm{ideal}-\partial X$ denote the orbifold obtained by adjoining to X its ideal boundary. Let $K_1 \subset K_2 \subset \dots$ denote an exhaustion of $\overline{X}$ by connected compact sets. Let γ denote a path in X and $\hat{\gamma}$ a lift to the universal cover Δ. Then the analogue of Lemma 2.1 becomes:

LEMMA 3.1. *There exists a sequence $\alpha(n) \to 0$ as $n \to \infty$, such that*

$$\mathrm{diam}(\hat{\gamma}) \leq \alpha(N(\gamma)),$$

where $N(\gamma)$ is the largest integer N such that γ is disjoint from K_N.

COROLLARY 3.2. *A homotopy to the identity rel ideal boundary is a homotopy rel $\mathbf{S}^1$ (and vice versa).*

PROOF: We mimic the proof of Theorem 2.2. Let ϕ_t be a homotopy rel ideal boundary, $\hat{\phi}_t$ the unique lift to a homotopy of $\Delta \cup \Omega$ such that $\hat{\phi}_0 = id$. We need only construct a neighborhood U of $\mathbf{S}^1$ such that $\mathrm{diam}(\hat{\phi}(I,z))$ is less than ϵ for all z in U.

Choose N such that $\alpha(N)$ is less than ϵ, and choose M such that K_M contains $\phi(I, K_N)$. Then $\mathrm{diam}(\hat{\phi}(I,z)) < \epsilon$ whenever $\pi(z)$ lies outside K_M.

Let $L \subset \Delta \cup \Omega$ be a compact set such that $\pi(L) = K_M$. Then

$$\hat{\phi}(I,\gamma L) = \gamma(\hat{\phi}(I,L)) \qquad \text{for all } \gamma \text{ in } \Gamma$$

and since $\hat{\phi}(I,L)$ is a compact subset of the domain of discontinuity, the Euclidean diameters of its translates under Γ tend to zero. Thus diam$(\hat{\phi}(I,z)) < \epsilon$ except possibly when z lies in one of finitely many translates of L. But these form a compact set, and the isotopy fixes $L \cap \mathbf{S}^1$, so there is a neighborhood of $\mathbf{S}^1$ on which diam$(\hat{\phi}(I,z)) < \epsilon$. $\square$

REMARK: These two results fail for hyperbolic three-manifolds. For example, take a totally degenerate limit of quasifuchsian groups, such that the domain of discontinuity Ω is simply connected and the quotient $S = \Omega/\Gamma$ is a compact Riemann surface. Then the quotient three-manifold plus ideal boundary $\overline{X}$ is homeomorphic to a product $S \times [0,\infty)$ (see Thurston [**Thur**] and Bonahon [**Bon**]). Thus the homotopy class of the closed geodesic corresponding to some fixed loxodromic element in the group is represented by loops arbitrarily far out in the end of $\overline{X}$. A path wrapping many times around such a loop has a lift which nearly connects the fixed points of the loxodromic element, so its diameter does not go to zero.

A similar argument gives an isotopy rel ideal boundary which does not lift to an isotopy rel $\mathbf{S}^2$. (By a straightforward extension of Proposition 2.3, such an isotopy cannot be uniformly quasiconformal.)

PROOF OF LEMMA 3.1: It suffices to show, for $\epsilon > 0$, there exists an N such that diam$(\hat{\gamma}) < \epsilon$ whenever γ lies outside K_N.

Let $\overline{X} = (\Delta \cup \Omega)/\Gamma$. The union of Ω and the hyperbolic fixed points of Γ is a dense subset of $\mathbf{S}^1$; choose a finite subset F which is ϵ-dense. Let L denote the hyperbolic convex hull of F. Then L/Γ is a compact subset of $\overline{X}$. (After taking the quotient, the parts of L near $\mathbf{S}^1$ either touch the ideal boundary or spiral around the closed geodesics corresponding to the hyperbolic fixed points).

Thus for N sufficiently large, $L/\Gamma \subset K_N$. The components of $\overline{\Delta} - L$ have diameter less than ϵ (remark: it is here the proof breaks down in higher dimensions). If γ lies outside K_N, its lift $\hat{\gamma}$ lies outside L, so its diameter is less than ϵ, completing the proof. $\square$

Clifford J. Earle, Cornell University, Ithaca, New York
and M.S.R.I., Berkeley, California

Curt McMullen, M.S.R.I., Berkeley, California
Present Address:
Princeton University, Princeton, New Jersey

154

REFERENCES

[Ahlf] Ahlfors, L., "Conformal Invariants," McGraw-Hill, 1973.

[AB] Ahlfors, L., and Bers, L., *Riemann mapping theorem for variable metrics*, Annals of Math **72** (1960), 385–404.

[B1] Bers, L., *Uniformization, moduli and Kelinian groups*, Bull. London Math. Soc. **4** (1972), 257–300.

[B2] Bers, L., *The moduli of Kleinian groups*, Russian Math Surveys **29** (1974), 88–102.

[B3] Bers, L., *On Sullivan's proof of the finiteness theorem and the eventual periodicity theorem*, Preprint.

[BG] Bers, L., and Greenberg, L., *Isomorphisms between Teichmüller spaces*, in *Advances in the Theory of Riemann Surfaces*, Princeton: Annals of Math Studies **66** (1971), 53–79.

[BR] Bers, L., and Royden, H.L., *Holomorphic families of injections*, Acta Math. **157** (1986), 259–286.

[Bon] Bonahon, F., *Bouts des varieties hyperbolique de dimension trois*, To appear.

[DE] Douady, A., and Earle, C., *Conformally natural extension of homeomorphisms of the circle*, Acta Math. **157** (1986), 23–48.

[DH] Douady, A., and Hubbard, J., *On the dynamics of polynomial-like mappings*, Ann. Sci. Ec. Norm. Sup. **18** (1985), 287–344.

[EE1] Earle, C., and Eells, J., *On the differential geometry of Teichmüller spaces*, J. Analyse Math. **19** (1967), 35–52.

[EE2] ________________, *A fibre bundle description of Teichmüller theory*, J. Diff. Geom. **3** (1969), 19–43.

[Eps] Epstein, D.B.A., *Curves on 2-manifolds and isotopies*, Acta Math. **115** (1966), 83–107.

[FRW] Fitzgerald, C.H., Rodin, B., and Warschawski, S.E., *Estimates for the harmonic measure of a continuum in the unit disk*, Trans. AMS 287 (1985), 681–685.

[Gar] Gardiner, F., *A theorem of Bers and Greenberg for infinite dimensional Teichmüller spaces*, These proceedings.

[Mar] Marden, A., *On homotopic mappings of Riemann surfaces*, Annals of Math. 90 (1969), 1–8.

[Sul] Sullivan, D., *Quasiconformal homeomorphisms and dynamics I: Solution of the Fatou-Julia problem on wandering domains*, Annals of Math. **122** (1985), 401–418.

[Thur] Thurston, W., *Geometry and Topology of Three Manifolds*, Princeton lecture notes.

The coefficient problem for univalent functions
with quasiconformal extension

BY S.L. KRUSHKAL'

§1. Introduction.

Univalent functions with quasiconformal extension are important in the theory of Teichmüller spaces as well as being interesting for their own sake. In this paper we obtain the exact bound for the coefficients of functions in the class $S(k)$ of normalized univalent functions on the unit disk with k-quasiconformal extension, where k is small. This answers a question of Kühnau and Niske [13]. The method is based on application of the known properties of extremal quasiconformal mappings and on the generalization of the Schwarz lemma.

§2. The Class $S(k)$.

Let S be the class of functions $f(z) = z + \sum_{n=2}^{\infty} a_n z^n$ univalent in the unit disk $\Delta = \{|z| < 1\}$. The class $S(k)$ consists of $f \in S$ admitting k-quasiconformal extension $\tilde{f}$ onto the Riemann sphere with the additional normalization $\tilde{f}(\infty) = \infty$. On $\Delta^* = \{|z| > 1\}$ we have the Beltrami coefficient $\mu_f = f_{\bar{z}}/f_z$ of the mapping $f = f^\mu \in S(k)$. Finally, we define the ball $B(\Delta^*) = \{\mu \in L^\infty(\mathbb{C}) : \mu \mid_\Delta = 0, \|\mu\|_\infty < 1\}$. For small k one has the formula for $S(k)$:

$$(1) \qquad f^\mu(\xi) = \xi - \frac{\xi^2}{\pi} \iint_{\Delta^*} \frac{\mu(z)\, dx\, dy}{z^2(z - \xi)} + O(\|\mu\|^2),$$

$z = x + iy$. From this one may easily deduce

$$(2) \qquad |a_n| \leq \frac{2k}{(n-1)} + O(k^2), \quad k \to 0,$$

see [7], Kühnau [12], and O. Lehto [15]. In fact there is the uniform bound

$$(3) \qquad |a_n| \leq \frac{2k}{n-1} + \left(n + \frac{2}{n-1}\right) k^2,$$

see [9]. Kühnau and Niske [13] asked whether there exists a $k(n) > 0$ such that for $f \in S(k)$, $k < k(n)$,

$$(4) \qquad |a_n| \leq \frac{2k}{(n-1)}.$$

This would be sharp. For $n = 2$ there is the well-known bound $|a_n| \leq 2k$ with equality only for the function $f_1(z) = \frac{z}{(1-tkz)^2}$, $|z| \leq 1$, $|t| = 1$. For $n \geq 3$ we would expect equality for

$$f_{n-1}(z) = \{f_1(z^{n-1})\}^{1/(n-1)} = z + \frac{2tk}{(n-1)}z^n + \cdots$$

The Beltrami coefficient for f_{n-1} may be taken to be $kt\mu_n$ where $|t| = 1$ and

$$(5) \qquad \mu_n(z) = \frac{|z|^{n+1}}{\bar{z}^{n+1}}$$

The previous best bound was

$$(6) \qquad |a_n| \leq \frac{2k}{(n-1)} + \left(\frac{2}{n-1} + n\right)k^4, \quad k \to 0,$$

see [9, 10].

§3. Statement of Results.

We consider a generalized extremal problem. Define a functional F of the form

$$F(f) = a_n + H(a_{m_1}, a_{m_2}, \ldots, a_{m_s})$$

where $a_j = a_j(f)$; $2 \leq n, m_k$ and H is a holomorphic function of s variables in an appropriate domain of $\mathbf{C}^s$. We assume that this domain contains the origin O and that $H, \partial H$ vanish at O. The main result of this paper is:

THEOREM. *For any functional of the above form there exists a $k(F) > 0$ such that for $k < k(F)$:*

$$(7) \qquad \max_{S(k)} |F(f)| = |F(f_{n-1})|$$

for some $|t| = 1$.

This immediately solves the problem of Kühnau and Niske.

§4. Proof of the Theorem.

We use the following notation. For a functional $L : S \to \mathbb{C}$ define

$$(8) \qquad \hat{L}(\mu) = L(f^\mu), \quad \mu \in B(\Delta^*).$$

Then if L is complex Gateaux differentiable $\hat{L}$ is a holomorphic functional on $B(\Delta^*)$. All our functionals have this property.

For $\mu \in L^\infty(\Delta^*)$, $\varphi \in L_1(\Delta^*)$ we write

$$(9) \qquad \langle \mu, \varphi \rangle = -\frac{1}{\pi} \iint_{\Delta^*} \mu\varphi \, dx \, dy.$$

Now let f_0 be any function in $S(k)$ maximizing $|F|$ over $S(k)$ (the existence of f_0 follows from compactness). Also let μ_0 be the dilatation of f_0. We may assume that μ_0 is extremal in its class, i.e.

$$\|\mu_0\|_\infty = \inf \left\{ \|\mu\|_\infty \le k : f^\mu \,|_\Delta = f_0 \,|_\Delta \right\}.$$

Suppose that $\mu_0 \neq kt\mu_n$ for some $|t| = 1$ (μ_n is defined by (5)). We show, that for small k, this leads to a contradiction. Our first lemma is:

LEMMA 1. For $2 \le p \neq n$, $\langle \mu_0, \frac{1}{z^{p+1}} \rangle = 0$.

This result was stated in [9] but due to its central role we shall prove it here.

First we note that (from (1)):

$$(10) \qquad \left\langle \mu_0, \frac{1}{z^{p+1}} \right\rangle = \lim_{\tau \to 0} \frac{a_p(f^{\tau\mu_0})}{\tau}.$$

Then by a compuatation, as k and $\tau \to 0$

$$\max \left\{ |\hat{F}(\mu)| : \|\mu\|_\infty \le \tau k \right\} = \frac{k\tau}{\pi} \iint_{\Delta^*} \frac{dx \, dy}{|z|^{n+1}} + O_n(k^2\tau^2),$$

$$(11) \qquad\qquad\qquad = |\hat{F}(\tau\mu_0)| + O_n(k^2\tau^2).$$

Consider the auxiliary functional

$$\hat{F}_p(\mu) = \hat{F}(\mu) + (p-1)\xi \left\langle \mu, \frac{1}{z^{p+1}} \right\rangle$$

where $p \neq n$ is fixed and $\xi \in \mathbb{C}$. Then, similar to (11),

$$(12) \qquad \max_{B(\Delta^*)_k} \left| \hat{F}_p(\mu) \right| = \frac{k}{\pi} \iint_{\Delta^*} \left| \frac{1}{z^{n+1}} + \frac{(p-1)\xi}{z^{p+1}} \right| \, dx \, dy + O_n(k^2)$$

and the remainder term estimate is independent of p.

Using the known properties of the norm

$$h_p(\xi) = \iint_{\Delta^*} \left| z^{-n-1} + (p-1)\xi z^{-p-1} \right| \, dx \, dy$$

following from the Royden [16] and Earle-Kra [3] lemmas, we obtain from (11), (12) that for small ξ there should be

$$(13) \qquad \max_{B(\Delta^*)_k} \left| \hat{F}_p(\mu) \right| = \max_{B(\Delta^*)_k} \left| \hat{F}(\mu) \right| + k o_p(\xi) + O_p(k^2\xi) + O_n(k^2).$$

Now fix k and consider the classes $S(\tau k), \tau < 1$. Then from (13) we have

$$(13') \qquad \max_{B(\Delta^*)_{\tau k}} \left| \hat{F}_p(\mu) \right| = \max_{B(\Delta^*)_{\tau k}} \left| \hat{F}(\mu) \right| + \tau o_p(\xi) + O_p(\tau^2\xi) + O_n(\tau^2).$$

On the other hand, as $\xi \to 0$, $\tau \to 0$,

$$|\hat{F}_p(\tau\mu_0)| = |\hat{F}(\tau\mu_0)| + \mathrm{Re} \, \frac{\overline{\hat{F}(\tau\mu_0)}}{|\hat{F}(\tau\mu_0)|}(p-1)\xi \left\langle \tau\mu_0, \frac{1}{z^{p+1}} \right\rangle + O(\tau^2\xi^2)$$

$$= |\hat{F}(\tau\mu_0)| + \tau(p-1) \, | \, \xi \, | \left| \left\langle \mu_0, \frac{1}{z^{p+1}} \right\rangle \right| + O(\tau^2\xi^2)$$

by suitable choices of $\xi \to 0$. Comparison with (13) implies $\left\langle \mu_0, \frac{1}{z^{p+1}} \right\rangle = 0$. Consider now the Grunsky coefficients of the function $\sqrt{f(z^2)}$, i.e.

$$\log \frac{(f(z^2))^{1/2} - (f(\xi^2))^{1/2}}{(z - \xi)} = \sum_{m,n=1}^{\infty} w_{m,n} z^m \xi^n,$$

taking the branch of logarithm which vanishes at 1. We are only interested in the diagonal terms of $w_{n-1}(f) = w_{(n-1),(n-1)}$. Then one can show that w_{n-1} is related to the Taylor coefficients of f by

$$(14) \qquad w_{n-1} = \frac{1}{2} a_n + P(a_2, \ldots, a_{n-1}),$$

where P is a polynomial without constant or linear terms (see [6]). Then it is known that for $f \in S$, $|w_{n-1}| \leq \frac{1}{(n-1)}$ with equality for functions f_{n-1} only. Hence from the general theorem about ranges of values of holomorphic functionals on classes of quasiconfomal mappings (see [11, p.1, Ch. II]), for $f \in S(k)$

$$|w_{n-1}(f)| \leq \frac{k}{(n-1)}$$

with equality only for the functions f_1.

We now consider the mapping $\Lambda_{n-1} : B(\Delta^*) \to B(\Delta^*)$ defined by

$$(15) \qquad \Lambda_{n-1}(\mu) = \{(n-1)w_{n-1}(\mu)\}\,\mu_n.$$

Let us compute the differential of Λ_{n-1} at $\mu = 0$. From (1) and (15) we see that this is an operator $P_n : L^\infty(\Delta^*) \to L^\infty(\Delta^*)$ given by

$$P_n(\mu) = \beta_n \langle \varphi_n, \mu \rangle \mu_n, \quad \varphi_n = \frac{1}{z^{n+1}}.$$

Let us define $P_n(\mu_0) = \alpha(k)\mu_n$. Now as clearly f_0 is not equivalent to $f_{n-1}(z)$,

$$\left\{\Lambda_{n-1}\left(\tfrac{t}{k}\mu_0\right) : |t| < 1\right\} \subsetneq \{|z| < 1\}.$$

Thus by Schwarz Lemma

$$(16) \qquad |\alpha(k)| < k.$$

Now consider the function

$$\nu_0 = \mu_0 - \alpha(k)\mu_n.$$

We show that ν_0 annihilates integrable analytic functions on Δ^*.

From Lemma 1 and the mutual orthogonality of the powers $z^m, m \in \mathbb{Z}$,

$$\left\langle \nu_0, \frac{1}{z^{p+1}} \right\rangle = 0$$

for $n \neq p = 2, 3, \ldots$ So we have only to show

$$\langle \nu_0, \varphi_n \rangle = 0, \quad \varphi_n = \frac{1}{z^{n+1}}.$$

Consider the conjugate operator

$$(17) \qquad P_n^*(\varphi) = \beta_n \langle \mu_n, \varphi \rangle \varphi_n$$

which maps $L_1(\Delta^*)$ to $L_1(\Delta^*)$. Then P_n^* fixes the subspace $\{\lambda\varphi_n : \lambda \in \mathbb{C}\}$. On the other hand, by definition, $P_n(\nu_0) = 0$. Thus for some λ

$$\langle \nu_0, \varphi_n \rangle = \lambda \langle \nu_0, P_n^*\varphi_n \rangle = \lambda \langle P_n\nu_0, \varphi_n \rangle = 0.$$

Consider now in $L_1(\Delta^*)$ the subspace $A(\Delta^*)$ of functions $\varphi(z)$ analytic in Δ^* which also satisfy the condition $\varphi(z) = O(|z|^{-3})$, as $|z| \to \infty$. Let us define

$$A(\Delta^*)^\perp = \{\mu \in L^\infty(\Delta^*) : \langle \mu, \varphi \rangle = 0 \text{ for all } \varphi \in A(\Delta^*)\}.$$

First we have the well-known:

160

LEMMA 2. *The functions* $\varphi_n(z) = \frac{1}{z^{n+1}}$, $n = 2, 3, \ldots$, *form a complete set in* $A(\Delta^*)$.

Consequently we have proved that $\nu_0 \in A(\Delta^*)^\perp$. We have assumed that μ_0 is the extremal dilatation for f_0. Then it follows from the well-known properties of extremal quasiconformal mappings (see, e.g. [5, 8, 11]) that

$$\|\mu_0\|_\infty = \inf\{\langle \mu_0, \varphi \rangle : \varphi \in A(\Delta^*), \|\varphi\|_1 \leq 1\}.$$

In general such a result is true if and only if μ is extremal for f^μ. Thus for any $\nu \in A(\Delta^*)^\perp$

$$\|\mu_0\|_\infty = \inf\{\langle \mu_0 + \nu, \varphi \rangle : \varphi \in A(\Delta^*), \|\varphi\|_1 \leq 1\} \leq \|\mu_0 + \nu\|_\infty.$$

Thus we have

LEMMA 3. *If* f_0 *is extremal*

$$(18) \qquad\qquad \|\mu_0\|_\infty = k \leq \|\mu_0 - \nu_0\|_\infty.$$

We may now prove the theorem. By (18) $k \leq \|\mu_0 - \nu_0\|_\infty = \|\alpha(k)\mu_n\|_\infty = |\alpha(k)|$ which contradicts (16). Hence f_0 is equivalent to f_{n-1} and we can take $\mu_0 = tk\mu_n$ for some $|t| = 1$.

§5. Uniform Estimates.

One might conjecture for the case of the coefficient functionals $F(f) = a_n(f)$ the $k(n)$ is given by $\{\frac{2}{n(n-1)}\}^{1/(n-2)}$. This would imply a somewhat stronger result that there is a single $k_0 > 0$ such that (4) holds uniformly for $f \in S(k)$, $k \leq k_0$. One might also be able to obtain a bound $|a_n| \leq (2 + c_2(k_0)k)k/(n-1)$ which would improve Göktürk's [4] estimate

$$|a_n| \leq A(k)kn^{-\frac{1}{2}-\alpha(k)}$$

where $A(k), -\alpha(k)$ increases as $[0,1)$, $\alpha(0) = 1/2$.

Institute of Mathematics, Siberian Division of the USSR Academy of Sciences
Novosibirsk, USSR

References

1. De Branges, L., *A proof of the Bieberbach conjecture*, Acta Math. **154, No. 1-2** (1985), 137–152.
2. Earle, C.J., *On holomorphic cross-sections in Teichmüller spaces*, Duke Math. J. **36, No. 2** (1969), 409–415.
3. Earle, C.J. and I. Kra, *On sections of some holomorphic families of closed Riemann surfaces*, Acta Math. **137, No 1-2** (1976), 49–79.
4. Göktürk, Z., *Estimates for univalent functions with quasiconformal extensions*, Ann. Acad. Sci. Fenn., Ser. AI **589** (1974), 1–21.
5. Hamilton, R.S., *Extremal quasiconformal mappings with prescribed boundary values*, Trans. Amer. Math. Soc. **138** (1969), 399–406.
6. Hummel, J.A., *The Grunsky coefficients of a schlicht function*, Proc. Amer. Math. Soc. **15, No. 1** (1964), 142–150.
7. Krushkal', S.L., *Some extremal problems for univalent analytic functions*, Doklady Acad. Nauk USSR **182, No. 4** (1968), 754–757. (Russian).
8. Krushkal', S.L., "Quasiconformal Mappings and Riemann Surfaces," V.H. Winston, Washington; John Wiley, New York, 1979.
9. Krushkal', S.L., *Asymptotic estimates and the properties of univalent functions with quasiconformal extension*, Siberian Math. J. **24, No. 3** (1983), 112–118. (Russian).
10. Krushkal', S.L., *Applications of multi-dimensional complex analysis to geometric function theory*, in "Complex Analysis and Applications (Proceedings of the International Conference on Complex Analysis and Applications, Varna 1983)," Sofia, 1985, pp. 133–140.
11. Krushkal', S.L., and R. Kühnau, "Quasikonforme Abbildungen — neue Methoden und Anwendungen," Teubner-Texte zur Math., Bd. 54, Teubner, Leipzig, 1983.
12. Kühnau, R., *Verzerrungssätze und Koeffizientenbedingungen vom Grunskyschen Typ für quasikonforme Abbildungen*, Math. Nachr. **48, No. 1-6** (1971), 77–105.
13. Kühnau, R., and W. Niske, *Abschätzung des dritten Koeffizienten bei den quasikonform fortsetzbaren schlichten Funktionen der Klasse S*, Math. Nachr. **78** (1977), 185–192.
14. Kühnau, R., and J. Timmel, *Asymptotische Koeffizientenabschätzungen für die quasikonform fortsetzbaren Abbildungen der Klasse S bzw. Σ*, Math. Nachr. **91** (1979), 357–362.
15. Lehto, O., *Quasiconformal mappings and singular integrals*, in "Symposia Mathematica XVIII," Academic Press, London–New York, 1976, pp. 429–453.
16. Royden, H.L., *Automorphisms and isometries of Teichmüller spaces*, in "Advances in the Theory of Riemann Surfaces," Ann. of Math. Stud., No. 66, Princeton University Press, Princeton, 1971, pp. 369–383.

Cone conditions and quasiconformal mappings

BY RAIMO NÄKKI AND BRUCE PALKA

1. Introduction.

Let f be a quasiconformal mapping of the open unit ball $B^n = \{x \in R^n : |x| < 1\}$ in euclidean n-space R^n onto a bounded domain D in that space. For dimension $n = 2$ the literature of geometric function theory abounds in results that correlate distinctive geometric properties of the domain D with special behavior, be it qualitative or quantitative, on the part of f or its inverse. There is a more modest, albeit growing, body of work that attempts to duplicate in dimensions three and above, where far fewer analytical tools are at a researcher's disposal, some of the successes achieved in the plane along such lines. In this paper we contribute to that higher dimensional theory some observations relating the behavior of f and f^{-1} to one of the venerable geometric conditions in analysis, the cone condition prominent in potential theory, geometric measure theory, and elsewhere. We first demonstrate that, when D obeys a specific interior cone condition along its boundary, f must satisfy a uniform Hölder condition in B^n. With regard to f^{-1}, the dual result one might anticipate – that an exterior cone condition satisfied by D at its boundary would lead to a uniform Hölder estimate for f^{-1} in D – is not, in general, true. We show, however, that in the presence of a certain auxiliary condition on D, one which is implied by an exterior cone condition when D is a Jordan domain in the plane, such a cone condition does exert a definite influence on the modulus of continuity of f^{-1}. Our results generalize those of Lesley [6] for plane conformal mappings. (See also [10].) In fact, if $n = 2$ and D is a Jordan domain, they reduce in the conformal case precisely to Lesley's theorems.

2. Notation and terminology.

In all matters pertaining to notation and terminology we conform to the usage in the book of Väisälä [13], unless otherwise stipulated. For the convenience of the reader, however, we briefly review some of the notation that appears frequently in the sequel.

Part of this research was done while R. Näkki was visiting the University of Texas at Austin during 1985-86.

We denote by $B^n(x_o, r)$ the open ball of radius r centered at the point x_o in R^n, $n \geq 2$, and by $S^{n-1}(x_o, r)$ the boundary sphere of $B^n(x_o, r)$. We employ the abbreviations $B^n(r) = B^n(0, r)$, $B^n = B^n(1)$, $S^{n-1}(r) = S^{n-1}(0, r)$, $S^{n-1} = S^{n-1}(1)$. The $(n-1)$-dimensional measure of S^{n-1} is designated by ω_{n-1}. We make the added convention that $\omega_o = 2$.

The conformal modulus of a family Γ of curves in $\overline{R}^n = R^n \cup \{\infty\}$ is denoted by $M(\Gamma)$. For sets E, F, and G we employ the notation $\Delta(E, F : G)$ (respectively, $\Delta_o(E, F : G)$) to indicate the family of arcs (respectively, open arcs) which join E and F through G.

If f is a quasiconformal mapping between domains D and D', then $K_I(f)$ and $K_O(f)$ signify the inner and outer dilatations of f. Thus, for each family Γ of curves in D it is the case that

$$\frac{M(\Gamma)}{K_O(f)} \leq M\left[f(\Gamma)\right] \leq K_I(f)M(\Gamma) \ .$$

Let $0 < \alpha \leq 1$. A function $f : A \to R^n$, where A is a subset of R^n, belongs to $\mathrm{Lip}_\alpha(A)$, *the Lipschitz class of order α in A*, if f satisfies a *uniform Hölder condition with exponent α in A*: there exists a constant M such that

$$|f(x) - f(y)| \leq M|x - y|^\alpha$$

for all x and y in A. It is by now common knowledge among complex analysts that a quasiconformal mapping f of a subdomain D of R^n into R^n belongs to $\mathrm{Lip}_\alpha(A)$ for each compact subset A of D, where $\alpha = K_I(f)^{1/(1-n)}$, and that this Hölder exponent cannot, in general, be replaced with a larger one.

3. Modulus estimates.

Modulus considerations occupy a critical part in what is to follow. In this section we assemble some requisite background material on the subject.

The Grötzsch ring domain. For $0 < r < 1$, let

$$R_G(r) = B^n \setminus \{te_n : 0 \leq t \leq r\} \ ,$$

where $e_n = (0, 0, \dots, 1)$. The domain $R_G(r)$ is called the n-dimensional Grötzsch ring domain corresponding to r. We use $\mu_G(r)$ to designate the modulus of the family of open arcs that join the boundary components of $R_G(r)$ through $R_G(r)$. Then

$$(1) \qquad \frac{\omega_{n-1}}{\left(\log \frac{\lambda_n}{r}\right)^{n-1}} \leq \mu_G(r) \leq \frac{\omega_{n-1}}{\left(\log \frac{1}{r}\right)^{n-1}} \ ,$$

where $\lambda_n \geq 4$ is a constant depending only on n [3]. In particular, $\lambda_2 = 4$. On the other hand, it is also true that

$$(2) \qquad 2^{n-1} c_n \log \frac{1}{1-r} \leq \mu_G(r) \leq 2^{n-1} c_n \log \frac{8}{1-r} \,,$$

in which c_n is the constant from the "spherical cap inequality,"

$$(3) \qquad c_n = \omega_{n-2} \left[2 \int_0^{\pi/2} (\sin \theta)^{\frac{2-n}{n-1}} \, d\theta \right]^{1-n} \,.$$

(See [1], [2], and [13].) For instance, $c_2 = 2/\pi$.

The Teichmüller ring domain. The n-dimensional Teichmüller ring domain corresponding to $r > 0$ is the domain

$$R_T(r) = R^n \setminus \{ t e_n : -1 \leq t \leq 0 \quad \text{or} \quad r \leq t < \infty \}$$

and the modulus of the family of open arcs joining the boundary components of $R_T(r)$ through this domain is denoted by $\mu_T(r)$. It is known – see [3] – that

$$\mu_T(r) = 2^{1-n} \mu_G \left(\sqrt{\frac{1}{1+r}} \right) \,.$$

Referring to (1), we extract from this identity the bounds

$$(4) \qquad \frac{\omega_{n-1}}{\left[\log \lambda_n^2 (1+r) \right]^{n-1}} \leq \mu_T(r) \leq \frac{\omega_{n-1}}{\left[\log (1+r) \right]^{n-1}} \,.$$

Modulus estimates in cones. For $0 < \beta < 1$, let

$$C_\beta^n = \left\{ x = (x_1, \dots, x_n) : x_n > |x| \cos \beta \pi \right\} \,,$$

the open cone in R^n with vertex at the origin, with axis along the positive x_n-axis and with vertex angle $2\beta\pi$. For $r > 0$, we write

$$C_\beta^n(r) = C_\beta^n \cap B^n(r) \,.$$

Any image of $C_\beta^n(r)$ under an isometry of R^n taking the origin to the point x_o will be termed a *β-cone of radius r with vertex at x_o*.

We record information on two specific curve families associated with cones. The first of these is the family of all arcs joining $S^{n-1}(r)$ to $S^{n-1}(R)$ through $C_\beta^n(R)$. Here $0 < r < R$. We have

$$(5) \qquad M \left[\Delta \left(S^{n-1}(r), S^{n-1}(R) : C_\beta^n(R) \right) \right] = \frac{\omega_{n-2} \int_0^{\beta\pi} (\sin \theta)^{n-2} \, d\theta}{\left(\log \frac{R}{r} \right)^{n-1}} \,,$$

a fact which follows immediately from [**13**, p. 23] once it is observed that the $(n-1)$-dimensional measure of $S^{n-1} \cap C_\beta^n$ is given by

$$m_{n-1}(S^{n-1} \cap C_\beta^n) = \omega_{n-2} \int_0^{\beta\pi} (\sin\theta)^{n-2}\, d\theta \ .$$

A second curve family of special relevance for this paper is the family of open arcs joining the line segment $A = \{te_n : r \le t \le R\}$ to ∂C_β^n through C_β^n. Here we settle for the estimates

$$(6) \quad \omega_{n-2}q(\beta)^{1-n}\log\frac{R}{r} \le M\left[\Delta_o(A, \partial C_\beta^n : C_\beta^n)\right] \le \omega_{n-2}q(\beta)^{1-n}\log\frac{aR}{r},$$

where

$$(7) \qquad\qquad q(\beta) = \int_0^{\beta\pi} (\sin\theta)^{\frac{2-n}{n-1}}\, d\theta$$

and

$$(8) \qquad\qquad a = 4\exp\left[\frac{2\omega_{n-1}q(\beta)^{n-1}}{\omega_{n-2}(\log 2)^{n-1}}\right]\ .$$

The lower bound in (6) is a direct consequence of Corollary 2 in [**5**]. The upper bound is easily derived using that same corollary, after the observation is made that $\Delta_o(A, \partial C_\beta^n : C_\beta^n)$ is minorized by the union of three other curve families; namely, $\Delta_o(S^{n-1}(r/2), S^{n-1}(r) : R^n)$, $\Delta_o(S^{n-1}(R), S^{n-1}(2R) : R^n)$ and $\Delta_o(A^*, \partial C_\beta^n : B^n(2R) \setminus B^n(r/2))$, with $A^* = \{te_n : r/2 \le t \le 2R\}$.

4. Cone conditions and quasiconformal mappings.

A set A in R^n is said to obey an *interior* (respectively, *exterior*) *β-cone condition* if there is an $R > 0$ of which the following is true: for each x_o in ∂A the set A (respectively, $R^n \setminus A$) contains a β-cone of radius R with vertex at x_o. Such conditions arise in potential theory and geometric measure theory, where they serve as elementary means to insure that A (respectively, $R^n \setminus A$) is uniformly "thick" at its boundary points. Typically – and this is certainly the case for the situations we intend to investigate – each of these conditions entails a restriction on $\beta : 0 < \beta \le \frac{1}{2}$.

Suppose that f maps B^n quasiconformally onto a bounded domain D satisfying an interior cone condition. Does f belong to some Lipschitz class in B^n? If D is a Jordan domain, the response to this query is yes, as we will see shortly. If, however, we wish to accommodate non-Jordan domains D in the reply, but still to reach an affirmative conclusion, we are forced to make some adjustment in the cone condition. With this goal in mind, we incorporate one of several possible modified conditions in a definition. We declare that D satisfies an *omnilateral interior β-cone condition* if there exists an $R > 0$ such that every point x_o of ∂D has arbitrarily small open neighborhoods U with the ensuing feature: corresponding to each component V of $D \cap U$ the domain D contains a β-cone C of radius R with vertex at x_o which has its "tip" in V, meaning that $C \cap B^n(x_o, r)$ lies in V for small $r > 0$. Roughly speaking, this requires that any approach to x_o from D be tantamount to approach through a suitable cone. A moment's thought reveals that, if D is to enjoy this kind of property, D must be boundedly connected at its boundary; i.e., there must be an integer $N = N(n, \beta)$ for which the following holds: every point x_o of ∂D has arbitrarily small open neighborhoods U such that $U \cap D$ has at most N components. Of course, when D is a Jordan domain, the omnilateral cone condition reduces to the simple interior cone condition defined originally.

We are now prepared to state our first result.

THEOREM 1. *Let f be a quasiconformal mapping of B^n onto a bounded domain D. Suppose that D satisfies an omnilateral interior β-cone condition. Then f belongs to $\mathrm{Lip}_\alpha(B^n)$ for*

$$(9) \qquad \alpha = \frac{1}{K_O(f)} \left[\int_0^{\beta\pi} (\sin\theta)^{\frac{2-n}{n-1}}\, d\theta \Big/ \int_0^{\pi/2} (\sin\theta)^{\frac{2-n}{n-1}}\, d\theta \right]^{n-1}.$$

PROOF: Since D is boundedly connected at its boundary, f has an extension to a continuous, finite-to-one mapping of $\overline{B}^n$ onto $\overline{D}$ [7], [13]. We retain the notation f for the extended mapping. Fix a point x_o on S^{n-1}. Theorem 6 in [11] will allow us to complete the proof, once it is demonstrated that

$$(10) \qquad |f(x) - f(x_o)| \le M|x - x_o|^\alpha$$

for all x on the radial segment of B^n terminating at x_o, where α is prescribed by (9) and M is a constant not depending on x_o.

Suppose that $f(x_0)$ has exactly m preimages on S^{n-1}. From [7] we learn that $f(x_0)$ possesses an open neighborhood W having the following two properties: $W \cap D$ has exactly m components, each with $f(x_0)$ on its

168

boundary; as y tends to $f(x_0)$ through a component of $W \cap D$, $f^{-1}(y)$ approaches a preimage of $f(x_0)$ – a different preimage for each such component. In tandem with this fact, our cone condition permits us to select a β-cone C in D with vertex at $f(x_o)$ and with the property that $f^{-1}(y) \to x_o$ as $y \to f(x_o)$ through C. Moreover, C can be chosen to have radius $2R$, a number independent of x_o. In establishing (10) we are free to assume that $f(x_o) = 0$ and that $C = C_\beta^n(2R)$. These assumptions merely serve to simplify notational considerations. Define d in $(0,1)$ by

$$d = \max \left\{ |x| : \text{dist}\left[f(x), \partial D\right] \geq R \sin \beta\pi \right\} .$$

Taking heed of the fact that $\text{dist}(Re_n, \partial D) \geq R \sin \beta\pi$, we can be certain that $|f^{-1}(Re_n)| \leq d$. The constant d clearly does not depend on x_o. We write $A_o = \{te_n : 0 < t \leq R\}$. The main step in the verification of (10) is to produce a constant M_1 with the property that

$$(11) \qquad |f(x)| = |f(x) - f(x_o)| \leq M_1 |x - x_o|^\alpha$$

for all x belonging to the set $E \cup f^{-1}(A_o)$, where $E = \overline{B}^n(d)$.

Fix such an x. We first treat the case in which $f(x) = re_n$ with $0 < r < R$. For this, set $A = \{te_n : r \leq t \leq R\}$ and consider the curve family

$$\Gamma = \Delta_o\left(E \cup f^{-1}(A), S^{n-1} : B^n\right) .$$

Since $E \cup f^{-1}(A)$ is a continuum containing both 0 and x, the extremal property of the Grötzsch ring domain [3] and the estimate (2) validate the assertion that

$$(12) \qquad M(\Gamma) \geq \mu_G(|x|) \geq 2^{n-1} c_n \log \frac{1}{1 - |x|} \geq 2^{n-1} c_n \log \frac{1}{|x - x_o|} .$$

For $\Gamma' = f(\Gamma)$, the subadditivity of the modulus yields

$$(13) \qquad M(\Gamma') \leq M\left[\Delta_o\left(f(E), \partial D : D\right)\right] + M\left[\Delta_o(A, \partial D : D)\right] .$$

The first term on the right-hand side of (13) is just a constant – call it b – plainly independent of x_o. As to the family $\Delta_o(A, \partial D : D)$, it is minorized by $\Gamma_1 \cup \Gamma_2$:

$$\Gamma_1 = \Delta_o\left(A, S^{n-1}(2R) : B^n(2R)\right) \quad , \quad \Gamma_2 = \Delta_o(A, \partial C_\beta^n : C_\beta^n) .$$

For Γ_1 we have an obvious upper bound

$$M(\Gamma_1) \leq \omega_{n-1}(\log 2)^{1-n} \, ,$$

while by (6)

$$M(\Gamma_2) \leq \omega_{n-2}\, q(\beta)^{1-n} \log \frac{aR}{r} \, .$$

The upshot of these remarks is that

$$
\begin{aligned}
M(\Gamma') &\leq b + \omega_{n-1}(\log 2)^{1-n} + \omega_{n-2}\, q(\beta)^{1-n} \log \frac{aR}{r} \\[2mm]
&\leq \omega_{n-2}\, q(\beta)^{1-n} \log \frac{cR}{r} \, ,
\end{aligned}
$$

(14)

where

$$c = a \exp\left\{ \left[b + \omega_{n-1}(\log 2)^{1-n}\right] q(\beta)^{n-1} / \omega_{n-2} \right\} \, .$$

Because $M(\Gamma) \leq K_O(f) M(\Gamma')$, inequalities (12) and (14) lead to

$$\log \frac{r}{cR} \leq \frac{2^{n-1} q(\beta)^{n-1} c_n}{\omega_{n-2} K_O(f)} \log |x - x_o| = \alpha \log |x - x_o|$$

with α given by (9), as a glance at (3) confirms. Since $r = |f(x)|$, we arrive at the estimate

$$|f(x)| \leq cR |x - x_o|^\alpha$$

for any x on $f^{-1}(A_o)$. On the other hand, it is evident that

$$|f(x)| \leq \mathrm{dia}(D) \leq \frac{\mathrm{dia}(D)}{(1-d)^\alpha} |x - x_o|^\alpha$$

when x lies in E. We have thus established (11) with

$$M_1 = \max\left\{ cR,\, \mathrm{dia}(D)/(1-d)^\alpha \right\} \, .$$

Finally, to finish the verification of (10) we invoke Theorem 1 in [11], a Lindelöf-type theorem relating asymptotic and radial Hölder conditions for quasiconformal mappings. It insures the existence of a constant $M \geq M_1$ such that (10) holds for all x on the radial segment of B^n terminating at x_o. Furthermore, the proof of the theorem shows that M depends only on M_1, the dilatations of f, and the dimension n. In particular, it does not depend on x_o.

With inequality (10) in hand, we can appeal to Theorem 6 in [11] and assert that f belongs to $\mathrm{Lip}_\gamma(B^n)$ for $\gamma = \min\{\alpha, K_I(f)^{1/(1-n)}\}$. The fact that

$$\alpha \le \frac{1}{K_O(f)} \le K_I(f)^{1/(1-n)}$$

[13, p.48] means that f is a member of $\mathrm{Lip}_\alpha(B^n)$, as claimed.

REMARK 1: Suppose that in Theorem 1 we take $n = 2$. Then $K_O(f) = K_I(f) = K(f)$, the maximal dilatation of f. The Hölder exponent in (9) becomes simply $\alpha = 2\beta/K(f)$. For a conformal mapping f, therefore, $\alpha = 2\beta$. Accordingly, in the case where D is a plane Jordan domain and f is conformal, Theorem 1 reduces to a result of Lesley [6].

It is possible to recast Theorem 1 in the language of the theory of prime ends [8]. The generalization of the interior cone condition appropriate to the prime end setting is as follows: for each prime end P of D there exists in D a β-cone C of fixed radius R along which P is accessible from D – in the sense that, as x approaches the vertex of C through C, x tends to P in the topology of the prime end compactification of D. Under this assumption on D the conclusion of Theorem 1 follows without change. The proof requires slight alteration, for it is not immediately evident that the accessibility of prime ends through cones of fixed radius guarantees that f admits a continuous extension to $\overline{B}^n$. What results in [8] clearly show, however, is that f does possess a radial limit at each point of S^{n-1}. If in the proof of Theorem 1 we interpret $f(x_o)$ to mean the radial limit of f at x_o, the argument presented actually goes through to establish the prime end version of the theorem.

We next consider a quasiconformal mapping f of a bounded domain D onto B^n. An optimistic conjecture dual to Theorem 1 would be this: if D satisfies an exterior cone condition, then f is uniformly Hölder continuous in D. Such optimism turns out to be unwarranted. Indeed, save in dimension $n = 2$, even the assumption that D is a Jordan domain does not improve matters. Additional restrictions on D are imperative, if Hölder continuity is to be inferred. One pertinent constraint is a condition studied in [9].

We refer to a domain D in R^n as *b-arcwise connected* if each pair of points x and y in D can be joined in D by an arc γ satisfying $\mathrm{dia}(\gamma) \le b|x - y|$. Here $b \ge 1$ is a constant. A domain which displays for some b this generalization of the notion of convexity will be termed *quasiconvex*. Examples of these domains include, of course, convex domains. A much-studied class of quasiconvex domains is the class of *quasiballs* in R^n, i.e., the domains in R^n which are images of B^n under quasiconformal self-mappings of $\overline{R}^n$. If D is a bounded quasiconvex domain and if D is quasiconformally equivalent to B^n, then D is necessarily a Jordan domain, for any such domain is obviously locally connected at each of its boundary points. A bounded Jordan domain in R^2 that satisfies an exterior cone condition is quasiconvex, as results in [12] imply. (For details see [10].) In higher dimensions the corresponding assertion need not be true.

It was demonstrated in [9] that a quasiconformal mapping of a bounded quasiconvex domain D onto B^n belongs to $\mathrm{Lip}_\alpha(D)$ for $\alpha = \left[2K_I(f)\right]^{1/(1-n)}$. With the addition of an exterior cone condition this result can be sharpened appreciably.

THEOREM 2. *Let f be a quasiconformal mapping of a bounded quasiconvex domain D onto B^n. Suppose that D satisfies an exterior β-cone condition. Then f belongs to $\mathrm{Lip}_\alpha(D)$ for*

$$(15) \qquad \alpha = \frac{1}{\left[cK_I(f)\right]^{1/(n-1)}} \, ,$$

where

$$(16) \qquad c = 2 - \left[\int_0^{\beta\pi} (\sin\theta)^{n-2}\, d\theta \Big/ \int_0^{\pi/2} (\sin\theta)^{n-2}\, d\theta\right] .$$

PROOF: Since D is a Jordan domain, f admits an extension to a homeomorphism between $\overline{D}$ and $\overline{B}^n$, which extension we continue to call f. We show first that

$$(17) \qquad |f(x) - f(x_o)| \leq M|x - x_o|^\alpha$$

whenever x_o belongs to ∂D and x lies in D. Here α is the exponent specified by (15) and (16), and M is a constant independent of x_o and x. In doing so, we make use of the notation

$$A' = \overline{B}^n(\tfrac{1}{2}) \quad , \quad A = f^{-1}(A') \quad , \quad d = \mathrm{dist}(A, \partial D) .$$

We also choose R in $(0, d)$ such that the exterior β-cone condition is satisfied by D with cones of radius R.

Fix x_o on ∂D, x in D and write

$$r = |x - x_o| \quad , \quad s = |f(x) - f(x_o)| .$$

Assuming that D is b-arcwise connected, we initially treat the case in which $r < R/2b$. Our assumptions imply that x_o and x are the endpoints of an arc E lying, apart from the point x_o itself, in D and satisfying dia $(E) < 2br$. Let $\Gamma = \Delta(A, E : D)$. Then Γ is minorized by the curve family

$$\Gamma_1 = \Delta\big(S^{n-1}(x_o, R), S^{n-1}(x_o, 2br) : B^n(x_o, R)\backslash\overline{C}\big) \,,$$

where C is a β-cone of radius R in $R^n\backslash D$ with vertex at x_o. Since $B^n(x_o, R)\backslash\overline{C}$ is a $(1-\beta)$-cone of radius R and since $2\omega_{n-2}\int_0^{\pi/2}(\sin\theta)^{n-2}\,d\theta = \omega_{n-1}$, we conclude with the help of (5) that

$$(18) \qquad M(\Gamma) \leq M(\Gamma_1) = \frac{c\,\omega_{n-1}}{2\left(\log\frac{R}{2br}\right)^{n-1}} \,,$$

with c as in (16). Let $E' = f(E)$. Using A'' and E'' to designate the images of A' and E' under inversion in the sphere S^{n-1}, we observe that $R^n\backslash(A''\cup E')$ is a ring domain which separates the points $f(x_o)$ and $f(x)$ from the points $2f(x_o)$ and ∞. A symmetry principle for the modulus [4, p.13], in combination with the extremal property of the Teichmüller ring domain and estimate (4), justifies the computation

$$M\big[f(\Gamma)\big] = M\big[\Delta(A', E' : B^n)\big] \geq \tfrac{1}{2}M\big[\Delta(A'\cup A'', E'\cup E'' : \overline{R}^n)\big]$$

$$\geq \tfrac{1}{2}M\big[\Delta(A'', E' : \overline{R}^n)\big] \geq \tfrac{1}{2}\mu_T\left(\frac{|2f(x_o) - f(x_o)|}{|f(x) - f(x_o)|}\right)$$

$$= \tfrac{1}{2}\mu_T(1/s) \geq \frac{\omega_{n-1}}{2\left[\log\lambda_n^2\left(1 + \frac{1}{s}\right)\right]^{n-1}} \,.$$

Because $1 + (1/s) \leq 3/s$, we are able to deduce that

$$(19) \qquad M\big[f(\Gamma)\big] \geq \frac{\omega_{n-1}}{2\left(\log\frac{3\lambda_n^2}{s}\right)^{n-1}} \,.$$

Since $M\big[f(\Gamma)\big] \leq K_I(f)M(\Gamma)$, it can be inferred from (18) and (19) that, subject to the constraint $|x - x_o| < R/2b$,

$$|f(x) - f(x_o)| = s \leq Mr^\alpha = M|x - x_o|^\alpha \,,$$

with $M = 3\lambda_n^2(2b/R)^\alpha$. When $|x - x_o| \geq R/2b$, it is trivially the case that

$$|f(x) - f(x_o)| \leq 2 \leq 2\left(\frac{2b}{R}\right)^\alpha|x - x_o|^\alpha \leq M|x - x_o|^\alpha \,.$$

This establishes (17).

Finally, Theorem 1 in [9] places f in $\mathrm{Lip}_\gamma(D)$ for $\gamma = \min\{\alpha, K_I(f)^{1/(1-n)}\}$. However, $1 \le c \le 2$, with the result that $\alpha \le K_I(f)^{1/(1-n)}$. Therefore, f is a member of $\mathrm{Lip}_\alpha(D)$.

REMARK 2: If we specialize Theorem 2 to the plane by letting $n = 2$, the Hölder exponent (15) simplifies to $\alpha = 1/[(2 - 2\beta)K(f)]$, with $K(f)$ the maximal dilatation of f. In the conformal case this becomes $\alpha = 1/(2-2\beta)$. Consequently, if D is a bounded plane Jordan domain that satisfies an exterior β-cone condition – as remarked earlier, D is then quasiconvex – and if f maps D conformally onto B^2, then f belongs to $\mathrm{Lip}_\alpha(D)$ for $\alpha = 1/(2 - 2\beta)$. This is just the content of Theorem 2 in [6].

Another interesting special case of Theorem 2 is captured in

COROLLARY 1. *Let f be a quasiconformal mapping of a bounded convex domain D onto B^n. Then f belongs to $\mathrm{Lip}_\alpha(D)$ for $\alpha = K_I(f)^{1/(1-n)}$.*

PROOF: A bounded convex domain is certainly quasiconvex and satisfies an exterior β-cone condition for $\beta = \frac{1}{2}$. Theorem 2 applies with $c = 1$.

REMARK 3: The Hölder exponent in Corollary 1 is optimal. To see this, just consider the radial stretching f of B^n onto itself defined by $f(x) = |x|^{\alpha-1}x$ for $x \ne 0$ and $f(0) = 0$, where $0 < \alpha \le 1$. Then $K_I(f) = \alpha^{1-n}$ and α is the largest possible Hölder exponent for f in B^n.

5. Disk conditions and quasiconformal mappings.

It should be apparent that the proofs of Theorems 1 and 2 do not rely on anything truly peculiar to cones to make them click. The results lend themselves to variations, wherein the role of the cone is played by some different "comparison domain," typically by some quasiball in which modulus estimates are easily carried out. One case in point is an actual round ball. Thus, we can speak of a bounded domain D as obeying an *interior* (respectively, *exterior*) *disk condition* if there exists $R > 0$ with the following property: corresponding to each point x_o of its boundary the set D (respectively, $R^n \backslash D$) contains an open ball B of radius R with x_o on ∂B. In analogy with the cone situation, we can also define what it means for D to satisfy an *omnilateral interior disk condition*; to wit, each boundary point x_o of D is to have arbitrarily small open neighborhoods U with this property: corresponding to each component V of $D \cap U$ the domain D contains an open ball B of fixed radius R such that x_o lies on ∂B and such that $B \cap B^n(x_o, r)$ is contained in V for small $r > 0$. It goes without saying that disk conditions, amounting essentially to "curvature conditions," are more restrictive than the corresponding cone conditions. In the context of this paper they lead, as might be expected, to improved Hölder exponents.

The facsimile of Theorem 1 involving a disk condition is:

THEOREM 3. *Let f be a quasiconformal mapping of B^n onto a bounded domain D. Suppose that D satisfies an omnilateral interior disk condition. Then f belongs to $\mathrm{Lip}_\alpha(B^n)$ for $\alpha = 1/K_O(f)$.*

PROOF: Proceeding exactly as we did in the proof of Theorem 1, we fix x_o on ∂B^n and demonstrate that an inequality of the type (10) is valid, now with $\alpha = 1/K_O(f)$. The proof follows verbatim the line of argument employed in establishing (10), except that here $d = \max\{|x| : \mathrm{dist}[f(x), \partial D] \geq R\}$ and the cone $C = C_\beta^n(2R)$ is replaced by a ball of radius R, which we may take to be $B = B^n(Re_n, R)$. The only significant change occurs in estimate (14), where a better upper bound can be derived for $M(\Gamma')$. Specifically, in the present situation the curve family $\Delta_o(A, \partial D : D)$ appearing in (13) is minorized by $\Delta_o(A, \partial B : B)$, with the consequence that by (2)

$$M\big[\Delta_o(A, \partial D : D)\big] \leq M\big[\Delta_o(A, \partial B : B)\big] = \mu_G\big[(R - r)/R\big]$$

$$\leq 2^{n-1} c_n \log \frac{8R}{r} \, .$$

This refined information leads, in turn, to the bound

$$M(\Gamma') \leq 2^{n-1} c_n \log \frac{a}{r} \, ,$$

with $a = 8R \exp(2^{1-n} b/c_n)$. The sharper modulus estimate translates directly into the improved Hölder exponent occurring in Theorem 3.

REMARK 4: The Hölder exponent α arising in Theorem 3 cannot, in general, be improved. The self-mapping f of B^n described in Remark 3 has $K_O(f) = 1/\alpha$ and α is the optimal Hölder exponent for f in B^n, a domain which obviously satisfies an omnilateral interior disk condition.

In the case of a plane conformal mapping f the Hölder exponent in Theorem 3 is just $\alpha = 1$. When D is a Jordan domain in the plane and f is conformal, this theorem restates a classical result often attributed to Kellogg.

Theorem 2 also has its counterpart in the present setting.

THEOREM 4. *Let f be a quasiconformal mapping of a bounded quasicon-vex domain D onto B^n. Suppose that D satisfies an exterior disk condition. Then f belongs to $\mathrm{Lip}_\alpha(D)$ for $\alpha = K_I(f)^{1/(1-n)}$.*

PROOF: Mimicking the proof of Theorem 2, we fix points x_o on ∂D and x in D. We need only derive an estimate $|f(x) - f(x_o)| \le M|x - x_o|^\alpha$, in which $\alpha = K_I(f)^{1/(1-n)}$ and M is a constant not depending on x_o or x. Notation is carried over from the earlier proof, although R now indicates a radius in $(0, d)$ for which the exterior disk condition is satisfied. With the exception of minor cosmetic changes, the derivation of the estimate sought differs from the derivation of inequality (17) in only one respect, that being the upper bound (18) obtained for $M(\Gamma)$. In the circumstances of the present theorem Γ is minorized by $\Gamma_1 = \Delta\big(S^{n-1}(x_o, R), S^{n-1}(x_o, 2br) : R^n \backslash \overline{B}\big)$. Here B is an open ball of radius R lying in $R^n \backslash D$ and having x_o on its boundary. We verify that, when $r = |x - x_o| < R/4b$,

$$(20) \qquad M(\Gamma_1) \le \frac{\omega_{n-1}}{2\left(\log \frac{R}{4br}\right)^{n-1}} \cdot$$

The consequence of (20) is a sharper bound for $M(\Gamma)$, which the proof then converts into the desired analogue of (17), with M now given by $M = 3\lambda_n^2(4b/R)^\alpha$.

In establishing (20) we may assume that $x_o = 0$ and that $B = B^n(Re_n, R)$. Let ψ designate the inversion in ∂B. Straightforward computations reveal that

$$|\psi(x)| \ge R/2$$

for all x in $S^{n-1}(R)\backslash B$, while

$$|\psi(x)| \le 2br$$

for all x in $S^{n-1}(2br)\backslash B$. Since $\psi(R^n \backslash \overline{B}) \subset B$ and $2br < R/2$, it is apparent that $\psi(\Gamma_1)$ is minorized by the family $\Gamma_2 = \Delta\big(S^{n-1}(R/2), S^{n-1}(2br) : C_\beta^n\big)$ with $\beta = \frac{1}{2}$. This insures that

$$M(\Gamma_1) = M\big[\psi(\Gamma_1)\big] \le M(\Gamma_2) = \frac{\omega_{n-1}}{2\left(\log \frac{R}{4br}\right)^{n-1}},$$

as claimed.

REMARK 5: Once again a radial stretching of B^n onto itself reveals that the Hölder exponent in Theorem 4 is not, in general, subject to improvement.

If D is a bounded Jordan domain in R^2 that satisfies an exterior disk condition, then D is automatically quasiconvex and so, by Theorem 4, any conformal mapping of D onto B^2 must belong to $\mathrm{Lip}_\alpha(D)$ for $\alpha = 1$. This is Theorem 14 in [10].

Raimo Näkki, Department of Mathematics, University of Jyväskylä, Jyväskylä, Finland

Bruce Palka, Department of Mathematics, University of Texas at Austin, Austin, TX 78712

REFERENCES

1. Anderson, G.D., *Extremal rings in n-space for fixed and varying n*, Ann. Acad. Sci. Fen. **AI 575** (1974), 1–21.
2. Caraman, P., "*n*-dimensional quasiconformal (QCf) mappings," Abacus Press, Tunbridge Wells, Kent, England, 1974.
3. Gehring, F.W., *Symmetrization of rings in space*, Trans. Amer. Math. Soc. **101** (1961), 499–519.
4. Gehring, F.W. and Väisälä, J., *The coefficients of quasiconformality of domains in space*, Acta Math. **114** (1965), 1–70.
5. Hag, K. and Vamanamurthy, M.K., *The coefficients of quasiconformality of cones in n-space*, Ann. Acad. Sci. Fenn. **AI 3** (1977), 267–275.
6. Lesley, F.D., *Conformal mappings of domains satisfying a wedge condition*, Proc. Amer. Math. Soc. **93** (1985), 483–488.
7. Näkki, R., *Boundary behavior of quasiconformal mappings in n-space*, Ann. Acad. Sci. Fenn. **AI 484** (1970), 1–50.
8. ________, *Prime ends and quasiconformal mappings*, J. d'Analyse Math. **35** (1979), 13–40.
9. Näkki, R. and Palka, B., *Lipschitz conditions and quasiconformal mappings*, Indiana Univ. Math. J., **31**, No. 3 (1982), 377–401.
10. ________________, *Extremal length and Hölder continuity of conformal mappings*, Comment. Math. Helv. **61** (1986), 389–414.
11. ________________, *Asymptotic values and Hölder continuity of quasiconformal mappings*, J. d'Analyse Math. (to appear).
12. Pommerenke, Ch., *One-sided smoothness conditions and conformal mapping*, J. London Math. Soc. **(2) 26** (1982), 77–88.
13. Väisälä, J., "Lectures on *n*-dimensional quasiconformal mappings," Lecture Notes in Mathematics **229**, Springer-Verlag, Berlin-Heidelberg-New York, 1971.

Existence of quasiregular mappings

BY SEPPO RICKMAN

1. Introduction.

Quasiregular mappings are roughly quasiconformal mappings without the homeomorphism requirement. In the Euclidean n-space R^n, $n \geq 2$, the definition is given as follows. A continuous map $f : G \to R^n$ of a domain G in R^n is called *quasiregular* (qr) if (1) f is in the local Sobolev space $W^1_{n,\mathrm{loc}}(G)$, i.e. f has weak first order partial derivatives which are locally in L^n, and (2) there exists K, $1 \leq K < \infty$, such that

$$(1.1) \qquad\qquad |f'(x)|^n \leq K J_f(x) \quad a.e.$$

Here $|f'(x)|$ is the operator norm of the formal derivative of f at x defined by means of the partial derivatives and $J_f(x)$ is the Jacobian determinant. For the purpose of this presentation f is also called K-quasiregular although it is not common terminology. The definition extends immediately to the case $f : M \to N$ where M and N are connected oriented Riemannian n-manifolds because both conditions (1) and (2) make sense also then. A map $f : M \to N$ is quasiconformal (qc) if and only if it is qr and a homeomorphism.

It has turned out that qr mappings form geometrically the right generalization of the one complex variable theory of analytic functions to real n-dimensional space. Many results in the classical 2-dimensional theory have their analogues for qr mappings in space. For example a Picard type theorem on omitted values and related results have been proved [10], [13], [14]. Even a much sharper result, namely, a defect relation in the spirit of Ahlfors' theory is true for quasiregular mappings [11].

I shall here consider the following general existence problem: Given two Riemannian n-manifolds M and N (all manifolds are assumed to be connected and oriented), does there exist a nonconstant qr mapping of M into N? The interesting case is when M is noncompact, and I am here almost completely concerned with the case $M = R^n$ for the simple reason that not very much is known so far for other manifolds M.

Let us now take a look at some basic facts about qr mappings. In the Euclidean case we have the following situation. For $n = 2$ the $1 - qr$

mappings are exactly the analytic functions, and a qr mapping f can always be written as a composition $g \circ h$ where h is qc and g analytic. For $n \geq 3$ the $1 - qr$ mappings are restrictions of Möbius transformations according to the generalized Liouville theorem. For this reason it is important to allow the factor K in (1.1).

Quasiregular mappings in their present form were introduced and studied first by Reshetnyak since 1966 in a series of papers. His results are contained in the book [9]. One of the main theorems by Reshetnyak is that a nonconstant qr mapping is open and discrete, i.e. point inverses are discrete sets. Note that this is exactly the topological description of analytic functions for $n = 2$. In 1967 Zorič [16] proved an important result about locally homeomorphic qr mappings for $n \geq 3$ which says that if $f : R^n \to R^n$ is such a map, then f is in fact a homeomorphism. This shows that, in a sense, interesting qr mappings for $n \geq 3$ have in general branching. In the same paper Zorič also posed the question of the validity of a Picard type theorem for qr mappings. For the basic theory of qr mappings I refer to [3], [4], [5], and [1].

2. Some examples and tools.

Many examples of qr mappings are obtained as finite-to-one branched coverings. One of the simplest of such maps is the winding map $f : R^3 \to R^3$, $f(r, \phi, x_3) = (r, k\phi, x_3)$ in cylinder coordinates where k is a positive integer.

Globally a more interesting example was given by Zorič in [16] as follows. It is a map $f : R^3 \to R^3 \backslash \{0\}$ which can be regarded as the counterpart of the exponential function in the plane. Let f_o be a qc mapping of the cylinder $C = \{x \in R^3 : 0 < x_1, x_2 < 1\}$ onto the half space $H_+ = \{x \in R^3 : x_3 > 0\}$ so that the edges of C correspond to rays emanating from the origin. We can extend f_o by repeated reflections through faces of C and ∂H_+ and obtain a qr mapping $f : R^3 \to R^3 \backslash \{0\}$. The branch set, i.e. the set where f fails to be a local homeomorphism, consists of the edges of C and their repeated reflections.

The most effective known tool in proofs in the theory of qr mappings is the method of extremal length, i.e. the use of certain inequalities of moduli of path families. If Γ is a family of (nonconstant) paths in M, the modulus is defined as

$$M(\Gamma) = \inf_{\rho} \int_M \rho^n \, dm$$

where the infimum is taken over all nonnegative Borel functions ρ whose line integral over all locally rectifiable paths $\gamma \in \Gamma$ is at least 1.

For a nonconstant $K - qr$ mapping $f : M \to N$ we have Poleckii's inequality [7]

$$(2.1) \qquad M(f\Gamma) \leq K^{n-1} M(\Gamma)$$

whenever Γ is a path family in M. The stronger inequality

$$(2.2) \qquad M(f\Gamma) \leq \frac{K^{n-1}}{m} M(\Gamma)$$

is due to Väisälä [15] and it holds when always at least m essentially separate paths in Γ are mapped onto one path.

For qc mappings the modulus is quasi-invariant but for qr mappings the converse of (2.1) need not hold. However, there are useful inequalities in that direction too which take the covering properties of f into account.

3.   The problem of Picard's theorem.

The question of the existence of a nonconstant qr mapping f of R^n into $R^n \backslash E$ with a given closed set E has been one of the main problems since the appearance of Zorič's paper [16]. At an early stage it was proved that E cannot have positive $(n-)$capacity [4], [8]. E has positive capacity if given any continuum C in $R^n \backslash E$ the family Γ of paths connecting C to E has positive modulus. We can give the proof of this fact with a simple application of (2.1) as follows. Let $f : R^n \to R^n \backslash E$ be a nonconstant $K - qr$ mapping and let $C = f\bar{B}^n$ be the image of the unit ball. Let Γ be as above and suppose $M(\Gamma) > 0$. Let Γ^* be the family of maximal lifts of the paths in Γ starting in $\bar{B}^n$. Then each path in Γ^* tends to ∞ and it is well known that then $M(\Gamma^*) = 0$. We also have $M(\Gamma) \leq M(f\Gamma^*) \leq K^{n-1} M(\Gamma^*) = 0$ which gives a contradiction.

For a long time it was conjectured that the Picard's theorem holds in space in the same form as in the plane, that is, card $E \geq 2$ would force any qr mapping $f : R^n \to R^n \backslash E$ to be constant. However, this conjecture turned out to be false and the solution of this problem of Picard's theorem is now known completely in a qualitative sense in dimension three. A Picard type theorem has been proved in [10] (see [12] and [13] for other proofs) for all dimensions in the following form.

3.1. THEOREM. *For every $K \geq 1$ there exists a positive integer $q = q(n, K)$ such that every $K - qr$ mapping $f : R^n \to R^n \backslash \{a_1, \ldots, a_q\}$ is constant.*

The solution for $n = 3$ is completed by

3.2 THEOREM [**14**]. *For every positive integer p there exists a nonconstant quasiregular mapping $f : R^3 \to R^3$ omitting p points.*

It seems that there is a possibility to obtain Theorem 3.2 also for $n \geq 4$ but this requires essential modifications to the proof given in [**14**].

4. Isoperimetric inequalities.

Isoperimetric inequalities stronger than the one true in R^n imply positive n-capacity at infinity. This fact has been pointed out and used by Gromov [**2**, Chapter VI]. See also the article [**6**] by Pansu.

Let $f : R^n \to N$ be quasiregular and let $\tilde{N}$ be the universal cover of the Riemannian n-manifold N. Suppose there exist numbers $c > 0$ and $m > n$ such that

$$(4.1) \qquad \operatorname{vol}(A) \leq cH^{n-1}(\partial A)^{m/(m-1)}$$

for all $A \subset\subset \tilde{N}$. Here H^k is the k-dimensional Hausdorff measure. Then $\tilde{N}$ has positive n-capacity at infinity, i.e. for any continuum $C \subset \tilde{N}$ the modulus $M(\Gamma)$ is positive where Γ is the family of paths connecting C to ∞ (see [**2**, Chapter VI]). If $\tilde{f} : R^n \to \tilde{N}$ is a lift of f, one shows as in the beginning of Section 3 that $\tilde{f}$, and hence f, must be constant.

4.2 EXAMPLES: Let us first take $N = S^1 \times S^2$. Then $\tilde{N}$ is quasiconformally equivalent to $R^3 \backslash \{0\}$. The Zorič example in Section 2 with the projection $\tilde{N} \to N$ gives then a qr mapping of R^3 into N.

If N is the connected sum $S^1 \times S^2 \sharp S^1 \times S^2$, $\pi_1 N$ is a free group of rank two and then $\tilde{N}$ satisfies (4.1) with $m = \infty$, i.e.

$$\operatorname{vol}(A) \leq cH^{n-1}(\partial A).$$

According to Sullivan's terminology, $\tilde{N}$ is open at infinity. Again by the capacity argument, there exists no nonconstant qr mapping $f : R^3 \to N$. This result follows also directly from Theorem 3.1 because $\tilde{N}$ is quasiconformally equivalent to $R^3 \backslash E$ where E is a Cantor set.

In the second example $\pi_1 N$ has exponential growth in the word metric. An interesting example of a compact manifold N with $\pi_1 N$ having polynomial growth is produced from the Heisenberg group G which is the Lie group of upper triangular matrices of the form

$$\begin{pmatrix} 1 & x & y \\ 0 & 1 & z \\ 0 & 0 & 1 \end{pmatrix}, \quad x, y, z \in R.$$

Let H be the subgroup of integer entries and let us fix a left invariant Riemannian metric on G. Then $N = G/H$ is a compact Riemannian 3-manifold with $\pi_1 N = H$ and $\tilde{N} = G$. G satisfies (4.1) with $m = 4$ (see [6]) and therefore there exists no qr mapping of R^3 into N. For the torus T^3 we have trivially the projection $R^3 \to T^3$ as an answer to the existence problem. The fundamental group $\pi_1 N$ grows polynomially but faster than $\pi_1 T^3$. This fact reflects the effect of the growth of the fundamental group to our problem. However, the proof of the isoperimetric inequality (4.1) with $m = 4$ for $\tilde{N} = G$ does not depend alone on the growth of $\pi_1 N$.

5. Comments and open problems.

Although some particular cases of the existence problem are quite well understood, like the Picard type theorems, the existence problem in more generality is to a large extent open. One would like to have general characterizations for example by means of homology and differential geometric concepts. I shall illustrate this by some examples.

In Section 4 all compact manifolds N had nontrivial $\pi_1 N$. If $N = S^2 \times S^2$, it is easy to give a qr mapping $f : R^4 \to N$ as follows. There exists a 2 to 1 branched covering $\phi : T^2 \to S^2$. We take f to be the projection $R^4 \to T^4 = T^2 \times T^2$ followed by $\phi \times \phi$. It is an open question whether there exists a nonconstant qr mapping of R^4 into $S^2 \times S^2 \sharp S^2 \times S^2$. It is also open whether R^4 can be mapped quasiregularly into $S^2 \times S^2 \backslash \{a\}$, or more generally, whether R^4 can be mapped quasiregularly into any $N_o \backslash \{a\}$ where N_o is a compact simply connected Riemannian n-manifold which is not covered by S^n. These comments show that when N is for example simply connected, the answer to our problem is not known even for very simple cases.

To show how homology plays a role, I shall first consider an example where N is compact. It can be shown that there is no nonconstant qr mapping of R^4 into $T^4 \sharp S^2 \times S^2$. In the proof both the nontriviality of

$\pi_1 T^4$ and $H_2(S^2 \times S^2)$ together with some other things are used, and the main argument is after some preparations similar to the one in the proof of Theorem 3.1. Although also $\pi_1(S^1 \times S^3)$ is nontrivial, a similar proof does not apply to maps of R^4 into $S^1 \times S^3 \sharp S^2 \times S^2$, and again, this case is open. Next consider $N = R^n \backslash \{a_1, \ldots, a_q\}$ with some Riemannian metric g. Here the codimension 1 homology has a dominant role. Namely, the right statement seems to be that there exists a number $q = q(n, K)$, independent of the metric g, such that every $K - qr$ mapping of R^n into (N, g) is constant. This Picard type statement and the generalization of 3.1 has not been proved in full generality yet. The difficulties are of technical nature and lie in the behavior of potentials for condensers.

For maps $f : M \to N$ with $M \neq R^n$ I shall make the following remarks. If $N = S^n$ and for some $L \geq 1$ M is given a triangulation each n-simplex of which can be mapped by an L-bilipschitz map followed by a homothety onto a standard simplex, then there exists a nonconstant qr mapping $f : M \to N$. We can construct f by taking a barycentric subdivision once and then using the well-known Alexander's construction.

Recently Gromov has pointed out a proof for the nonexistence of a nonconstant qr mapping of the Heisenberg group G into the hyperbolic space.

Department of Mathematics, University of Helsinki, 00100 Helsinki, Finland

185

REFERENCES

1. Bojarski, B. and Iwaniec, T., *Analytical foundations of the theory of quasiconformal mappings in R^n*, Ann. Acad. Sci. Fenn. Ser. AI Math. **8** (1983), 257–324.
2. Gromov, M., *Structures métriques pour les variétés riemanniennes*, Notes de cours rédigées par J. Lafontaine et P. Pansu, CEDIC-Fernand-Nathan, Paris, 1981.
3. Martio, O., Rickman, S. and Väisälä, J., *Definitions for quasiregular mappings*, Ann. Acad. Sci. Fenn. Ser. AI Math. **448** (1969), 1–40.
4. Martio, O., Rickman, S. and Väisälä, J., *Distortion and singularities of quasiregular mappings*, Ann. Acad. Sci. Fenn. Ser. AI Math. **465** (1970), 1–13.
5. Martio, O., Rickman, S. and Väisälä, J., *Topological and metric properties of quasiregular mappings*, Ann. Acad. Sci. Fenn. Ser. AI Math. **488** (1971), 1–31.
6. Pansu, P., *An isoperimetric inequality on the Heisenberg group*, Proceedings of "Differential Geometry on Homogeneous Spaces", Torino, 1983, 159–174.
7. Poleckii, E.A., *The modulus method for non-homeomorphic quasiconformal mappings (Russian)*, Mat. Sb. **83** (1970), 261–272.
8. Reshetnyak, J.G., *Extremal properties of mappings with bounded distortion (Russian)*, Sibirsk. Mat. Ž. **10** (1969), 1300–1310.
9. Reshetnyak, J.G., *Space Mappings with Bounded Distortion (Russian)*, Izdatel'stvo "Nauka" Sibirsk. Otdelenie, Novosibirsk, 1982.
10. Rickman, S., *On the number of omitted values of entire quasiregular mappings*, J. Analyse Math. **37** (1980), 100–117.
11. Rickman, S., *A defect relation for quasimeromorphic mappings*, Ann. of Math. **114** (1981), 165–191.
12. Rickman, S., *Value distribution of quasiregular mappings*, in "Value Distribution Theory: Proceedings of the Nordic Summer School in Mathematics, Joensuu, 1981," pp. 220–245, Lecture Notes in Mathematics **981**. Springer-Verlag, 1983.
13. Rickman, S., *Quasiregular mappings and metrics in the n-sphere with punctures*, Comment. Math. Helv. **59** (1984), 136–148.
14. Rickman, S., *The analogue of Picard's theorem for quasiregular mappings in dimension three*, Acta Math. **154** (1985), 195–242.
15. Väisälä, J., *Modulus and capacity inequalities for quasiregular mappings*, Ann. Acad. Sci. Fenn. Ser. AI Math. **509** (1972), 1–14.
16. Zorič, V.A., *The theorem of M.A. Lavrentjev on quasiconformal mappings in space (Russian)*, Mat. Sb. **74** (1967), 417–433.

Quasisymmetric maps

BY JUSSI VÄISÄLÄ

Introduction.

Quasisymmetric maps $f : R^1 \to R^1$ were introduced in 1956 in the famous paper [BA] of Beurling and Ahlfors, who proved that they are precisely the boundary maps of quasiconformal self homeomorphisms of the upper half plane fixing ∞. The present terminology was later suggested by Gehring and first published in the thesis of his student Kelingos.

In 1980 P. Tukia and the author extended the notion for maps between arbitrary metric spaces. Our main motivation was the need of a concept which could be used, for example, for maps between compact polyhedra or for embeddings $f : R^p \to R^n$, $p < n$, and which would share some properties with the quasiconformal maps. Later it turned out that these maps could also be used as a tool to prove results on quasiconformal maps. The related class of quasimöbius maps (see 1.5) seems to be still more useful for this purpose.

This article consists of two parts. The first part is a survey in which I give the definition, examples, and some properties of the quasisymmetric maps. In the second part I announce some new results on quasisymmetric invariants of closed sets in R^n.

1. Survey.

1.1. Notation. Throughout this article, X and Y will denote metric spaces, and the distance between points a and b will be written as $|a - b|$. We let $d(E)$ denote the diameter of a set. Open balls are written as $B(x, r)$, and closed balls as $\overline{B}(x, r)$.

1.2. Definition. A map $f : X \to Y$ is *quasisymmetric* (abbreviated QS) if (1) f is an embedding and (2) there is a homeomorphism $\eta : [0, \infty) \to [0, \infty)$ such that if x, a, b are distinct points in X, then

$$\frac{|f(a) - f(x)|}{|f(b) - f(x)|} \leq \eta \left(\frac{|a - x|}{|b - x|} \right).$$

We also say that f is $\eta - QS$.

1.3. Examples.

1. Every L-bilipschitz map is $\eta - QS$ with $\eta(t) = L^2 t$.

2. If $n \geq 2$, a map $f : R^n \to R^n$ is QS if and only if it is quasiconformal (abbreviated QC). More precisely, if f is $\eta - QS$, it is $K - QC$ with $K = \eta(1)^{n-1}$. Conversely, if f is $K - QC$, it is $\eta - QS$ with some η depending only on n and K [**Vä1**]. In fact, one can choose η to be independent on n [**AVV**]. However, there is an interesting open question: Can one choose η so that it depends only on the metric (or linear) dilatation

$$H(f) = \sup_{x \in R^n} \limsup_{r \to 0} \frac{L(x, f, r)}{l(x, f, r)}$$

where L and l are the maximum and minimum of $|f(x + h) - f(x)|$ for $|h| = r$? A positive answer would give some hope for the existence of a QC theory in infinite-dimensional spaces.

If G is a subdomain of R^n, an $\eta - QS$ map $f : G \to R^n$ is still $K - QC$ with $K = \eta(1)^{n-1}$, but a QC map need not be QS, because the boundary behvior of f easily destroys the quasisymmetry. In particular cases, for example, for maps between bounded uniform domains, the notions QS and QC are equivalent. Moreover, the notions locally QS and locally QC are equivalent for immersions $f : G \to R^n$.

3. A map $f : R^1 \to R^1$ is QS if and only if it is QS in the classical sense. This explains the terminology.

1.4. Discussion. The definition 1.2 seems to work well in all spaces. In several important cases it can be replaced by the following one: An embedding $f : X \to Y$ is *weakly* $H - QS$, $H \geq 1$, if $|a - x| \leq |b - x|$ implies $|f(a) - f(x)| \leq H|f(b) - f(x)|$. Obviously, QS implies weakly QS (with $H = \eta(1)$). The converse is true, for example, if $f : X \to R^n$ and if $X \subset R^n$ is of *bounded turning*, that is, there is $c \geq 1$ such that every pair a, b of points in X can be joined by a continuum $E \subset X$ with $d(E) \leq c|a - b|$. For maps $f : R^1 \to R^1$, this implies the equivalence of quasisymmetry and the Beurling–Ahlfors condition.

The quasisymmetry of a map is a *global* property while the quasiconformality is a *local* property. It seems to me that local conditions do not give interesting classes of maps between spaces of different dimensions.

1.5. Quasimöbius maps. If we replace in Definition 1.3 the ratio $|a - x|/|b - x|$ of three points by the cross ratio of four points, we obtain the definition of an η-*quasimöbius* map. This class includes both QS and möbius maps. Compared with the QS maps, these maps are more flexible in applications, but the theory is somewhat more complicated. See [**Vä2**].

1.6. Properties. I give a list of some results on QS maps. Proofs can be found in [**TV1**], [**TV2**] and [**Vä1**]. Suppose that $f : X \to Y$ is $\eta - QS$.

1. If X is bounded, then fX is bounded.

2. If X is complete, then fX is complete. In particular, if $Y = R^n$ and if X is closed in R^n, then fX is closed in R^n.

3. If X is connected, then f is $\eta_1 - QS$ with $\eta_1(t) = C \ \max(t^\alpha, t^{1/\alpha})$, where $C > 0$ and $\alpha \geq 1$ are constants depending only on η.

4. If X is bounded and connected, then f satisfies a Hölder condition

$$|x - y|^\alpha / C \leq |f(x) - f(y)| \leq C|x - y|^{1/\alpha},$$

where α depends only on η, and C depends only on $\eta, d(X)$ and $d(fX)$.

5. A topological circle $A \subset R^n$ is a QS circle if and only if it is of bounded turning.

6. If $n \geq 3$ and $1 \leq p \leq n - 1$, then R^n contains a QS p-cell which is not topologically flat in R^n. In this respect, QS maps differ from bilipschitz maps, since a bilipschitz p-cell in R^n is topologically flat whenever $p \leq n - 3$ [**LT**, 1.14].

7. If $X \subset R^n$ is either R^p or S^p, $1 \leq p \leq n - 1$, if $f : X \to R^n$ is $\eta - QS$ and if η is sufficiently close to the identity, then f can be extended to a $K - QC$ map $g : R^n \to R^n$ with K close to 1. For more general results of this type see [**Vä3**].

2. QS **invariants.**

2.1. Terminology. For a compact metric space X, we let $K(X)$ denote the space of all nonempty compact subsets of X, equipped with the Hausdorff metric. Then X is a compact metric space. We shall mainly consider the case where X is the extended n-space $\dot{R}^n = R^n \cup \{\infty\}$ with the spherical metric, and we write $K^n = K(\dot{R}^n)$. If $\infty \in A \in K^n$, a map $f : A \to R^n$ is called $\eta - QS$ if $f(\infty) = \infty$ and if $f|A \cap R^n$ is $\eta - QS$. (Such a map is always θ-quasimöbius with $\theta = \theta_\eta$.)

For every $H \subset K$ we set $\operatorname{sim} H = \{\phi A : A \in H, \phi \text{ similarity}\}$, $H^0 = \{A \in H : \{0, e_1\} \subset \partial A\}$. A family $H \subset K^n$ is called similarity invariant (SI) if $\operatorname{sim} H = H$. The concepts QS invariant (QSI) and topologically invariant (TI) are defined analogously. Thus H is QSI if $\phi A \in H$ whenever $A \in H$ and $\phi : A \to \dot{R}^n$ is QS. Trivially $TI \Rightarrow QSI \Rightarrow SI$.

2.2. Stable families. A family $L \subset K^n$ is called *stable* if (1) L is SI, (2) L^0 is compact. For example, the family of all closed disks and half planes is stable in K^2. On the other hand the set $A \subset R^2$ in the figure is not a member of any stable family $L \subset K^2$. Indeed, it is easy to find a sequence of similarities $\phi_j : R^2 \to R^2$ such that $\{0, e_1\} \subset \partial \phi_j A$ for all j and such that $\phi_j A \to \dot{R}^2$.

2.3. The operator σ. For every $H \subset K^n$ we let $\sigma(H)$ denote the union of all stable subfamilies of H. Alternatively, one can show that for SI families, $A \in \sigma(H)$ if and only if $\mathrm{cl}((\mathrm{sim}\ \{A\})^0) \subset H^0$.

2.4. Main theorem. If H is QSI, then $\sigma(H)$ is QSI.

Quasiproof. Let $\eta : [0, \infty) \to [0, \infty)$ be a homeomorphism, and let $F^n(\eta)$ be the set of all $\eta - QS$ maps $f : A \to R^n$ such that $\{0, e_1\} \subset A \in K^n$ and $f|\{0, e_1\} = \mathrm{id}$. By equicontinuity properties of QS maps and by a generalized version of Ascoli's theorem, one can show that $F^n(\eta)$ is compact in the topology induced by the Hausdorff metric of $K(\dot{R}^n \times \dot{R}^n)$. If $L \subset H$ is stable, the restriction $F_1 = F^n(\eta)|L^0$ is compact, and thus the family $M = \{fA : A \in L^0, f \in F^n(\eta)\}$ is compact. Let L_1 be the family of all images of all members of L under $\eta - QS$ maps into $\dot{R}^n$. Then $L_1^0 = M$, and hence L_1 is stable, which implies $L_1 \subset \sigma(H)$.

2.5. Corollary. If H is TI, then $\sigma(H)$ is QSI.

2.6. Remarks.

1. The idea of the proof of 2.4 is due to P. Tukia, who used it to prove the QS invariance of porous sets (see 2.9).

2. Given a QSI family H, there arises a *recognition problem*: How to characterize the members of $\sigma(H)$. I shall give solutions in three cases. In all of these, H is in fact TI.

3. It is possible to give a quantitative version of 2.4. Suppose that $H \subset K^n$ is QSI. We say that a function $c \mapsto L_c$, defined for $c \in [1, \infty)$, is

a *parametrization* of $\sigma(H)$ if

(1) $c \leq d$ implies $L_c \subset L_d \subset H$,

(2) L_c is stable for all c,

(3) every stable subfamily of H is contained in some L_c.

Then $\sigma(H) = \cup \{L_c : c \geq 1\}$. In all the applications given below, $\sigma(H)$ has a natural parametrization. The quasiproof above yields the following result:

2.7. Theorem. Suppose that $H \subset K^n$ is QSI and that $c \mapsto L_c$ is a parametrization of $\sigma(H)$. If $A \in L_c$ and if $f : A \to \dot{R}^n$ is $\eta - QS$, then $fA \in L_{c_1}$ with $c_1 = c_1(c, \eta, n)$.

2.8. First application. The family K^n is trivially TI. Hence $\sigma(K^n)$ is QSI. One can show that $A \in \sigma(K^n)$ if and only if there is $c \geq 1$ such that for every $x \in \partial A$ and for every $r \in (0, d(\partial A))$ there is $z \in \overline{B}(x, r)$ with $B(z, r/c) \cap A = \emptyset$. We say that such a set A is *boundary porous*. For example, the set in Figure 2.2 is not boundary porous.

2.9. Second application. Set $H_0 = \{A \in K^n : A \text{ has empty interior}\}$. Then H_0 is TI and hence $\sigma(H_0)$ is QSI. In fact, $\sigma(H_0) = H_0 \cap \sigma(K^n)$. A set A is in $\sigma(H_0)$ if and only if A is *porous* in R^n. By this we mean that there is $c \geq 1$ such that every ball $\overline{B}(x, r) \subset R^n$ contains a point z such that $B(z, r/c) \cap A = \emptyset$. From 2.7 it follows that if $A \subset R^n$ is c-porous and if $f : A \to R^n$ is $\eta - QS$, then fA is c_1-porous with $c_1 = c_1(c, \eta, n)$. By a different method [**Vä4**] I have proved that one can choose $c_1 = c_1(c, \eta)$. This method makes use of the topological degree, and it gives an explicit bound for $c_1(c, \eta)$.

2.10. Third application. Set $H_1 = \{A \in K^n : R^n \backslash A \text{ is connected}\}$, and assume $n \geq 2$. Then H_1 is TI, and thus $\sigma(H_1)$ is QSI. One can show that $A \in \sigma(H_1)$ if and only if $R^n \backslash A$ is a *uniform domain* in the sense of Martio and Sarvas [**MS**].

Department of Mathematics, University of Helsinki, Helsinki, Finland

REFERENCES

[**AVV**] Anderson, G.D., Vamanamurthy, M.K. and Vuorinen, M., *Dimension-free quasiconformal distortion in n-space*, Trans. Amer. Math. Soc. **297** (1986), 687–706.

[**BA**] Beurling, A. and Ahlfors, L., *The boundary correspondence under quasiconformal mappings*, Acta. Math. **96** (1956), 125–142.

[**LT**] Luukkainen, J. and Tukia, P., *Quasisymmetric and Lipschitz approximation of embeddings*, Ann. Acad. Sci. Fenn. Ser. AI Math. **6** (1981), 343–367.

[**MS**] Martio, O. and Sarvas, J., *Injectivity theorems in plane and space*, Ibid. **4** (1979), 383–401.

[**TV1**] Tukia, P. and Väisälä, J., *Quasisymmetric embeddings of metric spaces*, Ibid. **5** (1980), 97–114.

[**TV2**] ________________, *Extension of embeddings close to isometries or similarities*, Ibid. **9** (1984), 153–175.

[**Vä1**] Väisälä, J., *Quasi-symmetric embeddings in euclidean spaces*, Trans. Amer. Math. Soc. **264** (1981), 191–204.

[**Vä2**] _________, *Quasimöbius maps*, J. Analyse Math. **44** (1985), 218–234.

[**Vä3**] _________, *Bilipschitz and quasisymmetric extension properties*, Ann. Acad. Sci. Fenn. Ser. AI Math. **11** (1986), 239–274.

[**Vä4**] _________, *Porous sets and quasisymmetric maps*, Trans. Amer. Math. Soc. **299** (1987), 525–533.

A geometric interpretation of the Ahlfors–Weill mappings and an induced foliation of H^3

BY JOHN A. VELLING

1. Introduction.

In a 1962 paper [1] Ahlfors and Weill showed that a large class of univalent functions $F_\infty \colon \Delta \to \hat{\mathsf{C}} = \mathsf{C} \cup \infty$ can be extended to quasiconformal self-maps $A_\infty \colon \hat{\mathsf{C}} \to \hat{\mathsf{C}}$ as homeomorphisms of the entire sphere. The extension was obtained by illustrating quasiconformal homeomorphisms $F_0 \colon \Delta \to \hat{\mathsf{C}}$ whose images are precisely $\hat{\mathsf{C}} \backslash cl(F_\infty(\Delta))$. Both F_0 and F_∞ extend to $|z| = 1$ and agree there. Herein it will be shown that both F_∞ and F_0 are special cases of a more general collection of maps $F_t \colon \Delta \to cl(\mathsf{H}^3) = \mathsf{H}^3 \cup \hat{\mathsf{C}}$ where $\hat{\mathsf{C}} = \partial \mathsf{H}^3$. The maps F_t occur in a study of the ODE $g'' + Sg = 0$ on Δ and give rise to a foliation of $cl(\mathsf{H}^3)$ by 2-disks all agreeing on $|z| = 1$.

In fact if the map F_0 satisfies

$$\partial_{\bar{z}} F_0 = \mu_0 \partial_z F_0$$

then the quasiconformal distortion of each of the maps F_t is given by

$$\mu_t = \frac{1}{t^2 + 1} \mu_0$$

for $t \in [0, \infty]$. The Bers embedding of $\mathbf{T}_g \hookrightarrow \mathsf{C}^{3g-3}$ is given by assigning to a Riemann surface $\mathcal{R}$ with marking m, $(\mathcal{R}, m)$, the harmonic Beltrami differential associated with the class of marking preserving mappings from a fixed surface $(\mathcal{R}_0, m_0)$ considered as the origin of the Bers embedding. As harmonicity is preserved under multiplication by constants we have that rays in the Bers embedding occur as these foliations of $cl(\mathsf{H}^3)$.

This note is structured so that first given is a brief review of the Ahlfors–Weill paper. Then a map $M \colon PSU(1,1) = SU(1,1)/\pm id \to PSL(2,\mathsf{C}) = SL(2,\mathsf{C})/\pm id$ is given which factors as

$$
\begin{array}{ccc}
PSU(1,1) & \longrightarrow & PSL(2,\mathsf{C}) \\
SO(2) \Big\downarrow & \cdots\cdots\cdots\cdot\!> SO(3) \Big\downarrow \\
\Delta = \mathsf{H}^2 & \longrightarrow & \mathsf{H}^3
\end{array}
$$

194

for each $t \in (0, \infty)$. Next it is shown that these maps F_t foliate H^3 with boundary behavior given by F_0 and F_∞. Finally, some comments are made relating this to the Bers embedding of $\mathbf{T}_g$.

2. The Ahlfors–Weill Mappings.

Let S be a holomorphic function on Δ satisfying $\|S\|_\infty = \sup_{z \in \Delta} |S(z) \cdot (1 - z\bar{z})^2| < 1$. Let g_1 and g_2 be independent solutions of $g'' + Sg = 0$ on Δ. In the aforementioned Ahlfors–Weill paper it was shown that the maps

$$(1) \qquad F_\infty(z) = \frac{g_1(z)}{g_2(z)}$$

(so that the Schwarzian derivative $\{F_\infty, z\} = 2S$) and

$$(2) \qquad F_0(z) = \frac{\bar{z}g_1(z) + (1 - z\bar{z})g_1'(z)}{\bar{z}g_2(z) + (1 - z\bar{z})g_2'(z)}$$

agree on $|z| = 1$ in the sense that both are continuous on $|z| = 1$ and

$$\lim_{z \to e^{i\theta}} (F_0(z) - F_\infty(z)) = 0 \text{ for all } \theta \in [0, 2\pi).$$

It was also shown that F_0 and F_∞ map Δ to complementary regions on $\hat{\mathbb{C}}$ whose common boundary is a Jordan curve with zero area (Theorem A). As we can readily check that

$$(3) \qquad \partial_z F_0 = -S(z) \cdot (1 - z\bar{z})^2 \partial_{\bar{z}} F_0$$

we find that the map $A_\infty : \hat{\mathbb{C}} \to \hat{\mathbb{C}}$ given by

$$A_\infty(z) = \begin{cases} \overline{F_\infty(1/\bar{z})} & \text{if } z \in 1/\Delta \\ F_0(z) & \text{if } z \in \Delta \cup |z| = 1 \end{cases}$$

is a homeomorphism satisfying

$$\partial_{\bar{z}} A_\infty = \mu \partial_z A_\infty$$

where

$$\mu = \begin{cases} 0 & \text{if } z \in 1/\Delta \\ -\overline{S(z)} \cdot (1 - z\bar{z})^2 & \text{if } z \in \Delta. \end{cases}$$

3. The Monodromy Mapping.

We now consider this more carefully by closely examining the ODE $g'' + Sg = 0$ on Δ. The definition of the monodromy mapping found here is worked out in detail in another paper by the author [5]. For the sake of completeness, let z_α and z_β be two given disk coordinates on Δ, so that

$$z_\beta = e^{2i\theta} \frac{z_\alpha - p}{\overline{p}z_\alpha + 1} = \left[\frac{1}{(1 - p\overline{p})^{1/2}} \begin{pmatrix} e^{i\theta} & \\ & e^{-i\theta} \end{pmatrix} \begin{pmatrix} 1 & -p \\ -\overline{p} & 1 \end{pmatrix} \right] z_\alpha$$

or

$$z_\alpha = \left[\frac{1}{(1 - p\overline{p})^{1/2}} \begin{pmatrix} 1 & p \\ \overline{p} & 1 \end{pmatrix} \begin{pmatrix} e^{-i\theta} & \\ & e^{i\theta} \end{pmatrix} \right] z_\beta.$$

The bracketed notation is for the action of $PSU(1,1)$ on Δ as Möbius transformations. To change from α to β coordinates in the ODE it is classical that $g_\alpha'' + S_\alpha g_\alpha = 0$ transforms to $g_\beta'' + S_\beta g_\beta = 0$ if $S_\beta = S_\alpha \left(\frac{dz_\alpha}{dz_\beta} \right)^2$ and $g_\beta = g_\alpha \left(\frac{dz_\alpha}{dz_\beta} \right)^{-1/2}$. Thus S is a quadratic differential and g a $\left(-\frac{1}{2} \right)$ order differential on Δ. Denoting by $g_{\alpha,1}$ and $g_{\alpha,2}$ the solutions of $g_\alpha'' + S_\alpha g_\alpha = 0$ satisfying

$$\begin{pmatrix} g_{\alpha,1} & g_{\alpha,1}' \\ g_{\alpha,2} & g_{\alpha,2}' \end{pmatrix}_{z_\alpha = 0} = id$$

we may compare $\begin{pmatrix} g_{\alpha,1} \\ g_{\alpha,2} \end{pmatrix}$ with $\begin{pmatrix} g_{\beta,1} \\ g_{\beta,2} \end{pmatrix}$. As is shown in [5], we find that

$$\begin{pmatrix} g_{\alpha,1} \\ g_{\alpha,2} \end{pmatrix} = \frac{1}{(1 - p\overline{p})^{1/2}} \begin{pmatrix} g_{\alpha,1}(p) & \overline{p}g_{\alpha,1}(p) + (1 - p\overline{p})g_{\alpha,1}'(p) \\ g_{\alpha,2}(p) & \overline{p}g_{\alpha,2}(p) + (1 - p\overline{p})g_{\alpha,2}'(p) \end{pmatrix} \cdot$$

$$\cdot \begin{pmatrix} e^{i\theta} & \\ & e^{-i\theta} \end{pmatrix} \cdot \begin{pmatrix} g_{\beta,1} \\ g_{\beta,2} \end{pmatrix} \cdot \left(\frac{dz_\beta}{dz_\alpha} \right)^{-1/2}.$$

Given this, the monodromy mapping $M_S : PSU(1,1) \to PSL(2,\mathbb{C})$ is defined by

$$(4) \qquad M_S \left[\frac{1}{(1 - p\overline{p})^{1/2}} \begin{pmatrix} 1 & p \\ \overline{p} & 1 \end{pmatrix} \begin{pmatrix} e^{-i\theta} & \\ & e^{i\theta} \end{pmatrix} \right]$$

$$= \frac{1}{(1 - p\overline{p})^{1/2}} \begin{pmatrix} g_{\alpha,1}(p) & \overline{p}g_{\alpha,1}(p) + (1 - p\overline{p})g_{\alpha,1}'(p) \\ g_{\alpha,2}(p) & \overline{p}g_{\alpha,2}(p) + (1 - p\overline{p})g_{\alpha,2}'(p) \end{pmatrix} \begin{pmatrix} e^{i\theta} & \\ & e^{-i\theta} \end{pmatrix}$$

EXAMPLE: If $S = 0$ then $g_{\alpha,1} = 1$, $g_{\alpha,2} = z_\alpha$ and

$$M_S \begin{pmatrix} a & b \\ \overline{b} & \overline{a} \end{pmatrix} = \begin{pmatrix} \overline{a} & \overline{b} \\ b & a \end{pmatrix} \qquad \text{where} \qquad a\overline{a} - b\overline{b} = 1.$$

Now $PSL(2,\mathbb{C})$ acts on H^3 as the space of orientation preserving isometries. If H^3 is considered to have its standard upper half-space model coordinatized by $(z,t) \in \mathbb{C} \times \mathbb{R}_+$, with $\hat{\mathbb{C}} = \partial \mathsf{H}^3$, this action is given by

$$\begin{pmatrix} a & b \\ c & d \end{pmatrix} \cdot (z,t) = \left(\frac{(az+b)\overline{(cz+d)} + a\bar{c}t^2}{|cz+d|^2 + |c|^2 t^2} \;,\; \frac{t}{|cz+d|^2 + |c|^2 t^2} \right).$$

This action extends to $\hat{\mathbb{C}}$ and when restricted thereto is the action of Möbius transformations. So by means of M_S we can consider the action of $PSU(1,1)$ on H^3 by

$$\gamma : (z,t) \to M_S(\gamma) \cdot (z,t) \quad \text{for} \quad \gamma \in PSU(1,1).$$

Furthermore, as $\begin{pmatrix} e^{i\theta} & \\ & e^{-i\theta} \end{pmatrix}$ stabilizes $(0,t)$ for each $t \in \mathbb{R}_+$, $\theta \in [0, 2\pi)$ we have $F_{t,s} : \Delta \to \mathsf{H}^3$ defined by (dropping the subscript α)

$$F_{t,s}(p) = M_S \begin{pmatrix} 1 & p \\ \bar{p} & 1 \end{pmatrix} \cdot (0,t) =$$

(5)
$$\left(\frac{(\bar{p}g_1(p) + (1-p\bar{p})g_1'(p))\overline{(\bar{p}g_2(p) + (1-p\bar{p})g_2'(p))} + g_1(p)\overline{g_2(p)}t^2}{\text{denom}} \;,\; \frac{(1-p\bar{p})t}{\text{denom}} \right)$$

where denom $= |\bar{p}g_2(p) + (1-p\bar{p})g_2'(p)|^2 + |g_2(p)|^2 t^2$.

Let $\mathrm{stab}(\mathbf{x})$ denote the stabilizing group of $\mathbf{x}$ as a subgroup of the isometries of the appropriate ambient space. Since $\mathrm{stab}(0) \cong SO(2)$ for $0 \in \Delta$, and for any $\mathbf{x} \in \mathsf{H}^3$ we have $\mathrm{stab}(\mathbf{x}) \cong SO(3)$, we have for each $t \in \mathbb{R}_+$ a factoring of the diagram

$$
\begin{array}{ccc}
PSU(1,1) & \xrightarrow{\;M_S\;} & PSL(2,\mathbb{C}) \\[4pt]
\mathrm{stab}(0) \downarrow - \cdots \cdots \rightarrow \mathrm{stab}(0,t) \downarrow & & \\[4pt]
\Delta & \xrightarrow{\;F_{t,S}\;} & \mathsf{H}^3
\end{array}
$$

as mentioned in the introduction.

We also note that if $t = 0$ we get the orbit of $z = 0$ under the image $M_S(PSU(1,1))$ is given by

$$F_{0,S}(p) = \frac{\bar{p}g_1(p) + (1-p\bar{p})g_1'(p)}{\bar{p}g_2(p) + (1-p\bar{p})g_2'(p)}.$$

Similarly if $t = \infty$, or $z = \infty$ and $t = 0$ we get

$$F_{\infty,S}(p) = \frac{g_1(p)}{g_2(p)},$$

i.e. precisely the maps from the Ahlfors–Weill construction in (1) and (2).

4. A Foliation of H^3.

To establish the smoothness of these sheets we will need the following evident consequence of the Ahlfors–Weill paper:

COROLLARY A. *If g_2 is the solution of $g'' + Sg = 0$ where $\|S\|_\infty < 1$ such that $g_2(0) = 0$, $g_2'(0) = 1$, then*

$$|\bar{z}g_2(z) + (1 - z\bar{z})g_2'(z)| \neq 0$$

for all $z \in \Delta$.

We are now in a position to establish

THEOREM 1. *For S holomorphic on Δ so that $\|S\|_\infty < 1$ we have $F_{t,S} : \Delta \to \mathsf{H}^3$ is an immersion. If w_t is a conformal coordinate on $F_{t,S}(\Delta)$ then the quasiconformal distortion is given by*

$$(6) \qquad \mu_{t,S}(p) = \partial_{\bar{p}} w_t / \partial_p w_t = -\frac{\overline{S(p)} \cdot (1 - p\bar{p})^2}{(t^2 + 1)}.$$

PROOF: Let $\Phi(p, \bar{p}) = (Z, T)$ be the expression given in (5). It will be convenient to use $Z = X + iY$ or $\Phi = (X, Y, T)$. Let $p = u + iv$. As

$$\partial_p \Phi = \Phi_p = (X_p, Y_p, T_p)$$

$$= \frac{1}{2}(X_u - iX_v, Y_u - iY_v, T_u - iT_v)$$

and

$$\Phi_{\bar{p}} = \frac{1}{2}(X_u + iX_v, Y_u + iY_v, T_u + iT_v)$$

we have

$$\Phi_p \times \Phi_{\bar{p}} = (Y_p T_{\bar{p}} - Y_{\bar{p}} T_p, T_p X_{\bar{p}} - T_{\bar{p}} X_p, X_p Y_{\bar{p}} - X_{\bar{p}} Y_p)$$

$$= (i(T_p Z_{\bar{p}} - T_{\bar{p}} Z_p), \frac{i}{2}(Z_p \overline{Z}_{\bar{p}} - Z_{\bar{p}} \overline{Z}_p))$$

$$= \frac{i}{2}(X_u, Y_u, T_u) \times (X_v, Y_v, T_v)$$

with $A \times B$ considered as for vectors in $\mathbb{R}^3$. For Φ to be an immersion we thus need to show that $\Phi_p \times \Phi_{\bar{p}} \neq 0$ for any $p \in \Delta$.

Thanks is due here to Professor Joseph Oliger at Stanford University. With the aid of his **Navajo** Vax computer and the computational package **Maple**, various expressions were computed for the map Φ. Of course this could have been done by hand, and the computations were checked occasionally for correctness, but their tedious nature made them a candidate for the machine. It was found that

$$\Phi_p = \frac{1}{(\mathrm{denom})^2} \big([-t^2(t^2+1) - Sp^2(1-p\bar{p})^2]\bar{g}_2^2 - 2Sp(1-p\bar{p})^3\bar{g}_2\bar{g}_2' - S(1-p\bar{p})^4\bar{g}_2'^2 ,$$

$$[-t(t^2+1)\bar{p} + Spt(1-p\bar{p})^2]g_2\bar{g}_2 + tS(1-p\bar{p})^3 g_2\bar{g}_2' - t(t^2+1)(1-p\bar{p})g_2'\bar{g}_2' \big)$$

$$\Phi_{\bar{p}} = \frac{1}{(\mathrm{denom})^2} \big([(1+t^2)p^2 + \bar{S}t^2(1-p\bar{p})^2]\bar{g}_2^2 + 2p(t^2+1)(1-p\bar{p})\bar{g}_2\bar{g}_2'$$

$$+ (t^2+1)(1-p\bar{p})^2\bar{g}_2'^2 ,$$

$$[-t(t^2+1)p + \bar{S}pt(1-p\bar{p})^2]g_2\bar{g}_2 + t\bar{S}(1-p\bar{p})^3\bar{g}_2 g_2' - t(t^2+1)(1-p\bar{p})g_2\bar{g}_2' \big).$$

If (Z_1, T_1) is a tangent vector to H^3 at (Z, T) we have $|(Z_1, T_1)| = \frac{1}{T}\sqrt{|Z_1|^2 + T_1^2}$ so that

$$|\Phi_p \times \Phi_{\bar{p}}| = \frac{|(t^2+1)^2 - S\bar{S}(1-p\bar{p})^4|}{2t(1-p\bar{p})(\mathrm{denom})}.$$

We conclude that if $\|S\|_\infty < 1$ then each $F_{t,S}$ is an immersion of Δ. The metric on our surface in H^3 (for fixed t) is given by

$$\mathbf{Met}(Z, T) = \frac{1}{T^2}\begin{pmatrix} \Phi_u \cdot \Phi_u & \Phi_u \cdot \Phi_v \\ \Phi_u \cdot \Phi_v & \Phi_v \cdot \Phi_v \end{pmatrix} = \frac{1}{T^2(\mathrm{denom})^4}\begin{pmatrix} E & F \\ F & G \end{pmatrix}.$$

The product here is the standard one on $\mathbb{R}^3$. Let $g = EG - F^2$. Using the complex coordinates on H^3 we define the "twist" product

$$\mathbf{Tw}(Z, T) = \begin{pmatrix} \langle\langle \Phi_p, \overline{\Phi_{\bar{p}}} \rangle\rangle & \langle\langle \Phi_p, \overline{\Phi_p} \rangle\rangle \\ \langle\langle \Phi_{\bar{p}}, \overline{\Phi_{\bar{p}}} \rangle\rangle & \langle\langle \Phi_{\bar{p}}, \overline{\Phi_p} \rangle\rangle \end{pmatrix} = \frac{1}{(\mathrm{denom})^4}\begin{pmatrix} A & C \\ \overline{C} & B \end{pmatrix}$$

where $\langle\langle (Z_1, T_1), (Z_2, T_2) \rangle\rangle = Z_1 Z_2 + T_1 T_2$. These two are clearly related and $\mathbf{Tw}$ is computed because of its convenience in our notational system. In fact one readily checks that

$$E = A + B + C + \overline{C}, \quad F = \frac{\overline{C} - C}{i}, \quad G = A + B - C - \overline{C} \quad \text{and}$$

$$g = (A + B)^2 - 4C\overline{C}.$$

If one has a surface with u and v as given coordinates, the Gauss equations for isothermal coordinates in terms of a given metric on the surface,

$ds^2 = E\,dx^2 + 2F\,ds\,dy + G\,dy^2$, are

$$d\xi = \frac{1}{\lambda\sqrt{E}}(E\,du + F\,dv)$$

$$d\eta = \frac{\sqrt{g}}{\lambda\sqrt{E}}\,dv$$

as is well-known. From this we check that if $w_t = \xi + i\eta$ is a complex surface coordinate in isothermal parameters, and $p = u + iv$ is our coordinate from the disk via the immersion F_t, then the quasiconformal deformation is given by

$$\mu_{t,S} = (\partial_{\bar{p}} w_t)/(\partial_p w_t)$$

$$= \frac{\partial_u \xi - \partial_v \eta + i(\partial_u \eta + \partial_v \xi)}{\partial_u \xi + \partial_v \eta + i(\partial_u \eta + \partial_v \xi)}$$

$$= \frac{E - \sqrt{g} + iF}{E + \sqrt{g} - iF}$$

$$\tag{7} = \frac{E - G + 2iF}{E + 2\sqrt{g} + G}$$

$$= \frac{2\overline{C}}{(A + B) + \sqrt{(A + B)^2 - 4C\overline{C}}}$$

$$= -\frac{\overline{S} \cdot (1 - p\bar{p})^2}{(t^2 + 1)}$$

after some rather gruesome algebra. This concludes the proof.

NOTE 1: Equation (7) remains valid even if $\|S\|_\infty \geq 1$, but only for points $p \in \Delta$ where $|S(p)|^2(1 - p\bar{p})^4 < (t^2 + 1)^2$, which is true here by assumption.

NOTE 2: This is consistent with (3) as F_0 computed there is orientation reversing. Its Jacobian is

$$|\partial_p F_0|^2 - |\partial_{\bar{p}} F_0|^2 = \frac{|S(p) \cdot (1 - p\bar{p})^2|^2 - 1}{|\bar{p}g_2(p) + (1 - p\bar{p})g_2'(p)|^4}$$

which is well defined on Δ by corollary A.

Let Γ be a cofinite Fuchsian group, i.e. Δ/Γ has finite hyperbolic area. Let S be an analytic automorphic form of weight -4 with respect to Γ. In other words

$$\tag{8} S(\gamma z)\left(\frac{d\gamma z}{dz}\right)^2 = S(z)$$

for all $\gamma \in \Gamma$, $z \in \Delta$. Then $|S(z) \cdot (1 - z\bar{z})^2|$ is invariant under Γ and $\|S\|_\infty = \sup_{z \in \Delta/\Gamma} |S(z) \cdot (1 - z\bar{z})^2|$. Furthermore, $M_S(\Gamma)$ is a subgroup of $PSL(2, \mathbb{C})$, called the monodromy group of Γ under S. See [**2**, and **4**,§5.7] for example.

We now have

THEOREM 2. *If* $\|S\|_\infty < 1$ *then the surfaces* $F_{t,S}(\Delta)$ *foliate* H^3 *and agree with* $F_{0,S}$ *and* $F_{\infty,S}$ *on* $\partial\Delta$.

PROOF: This will follow if it can be shown that the surfaces are embedded and distinct (for distinct values of t) in H^3.

If $\|S\|_\infty < 1$ then by Ahlfors–Weill we have that both $F_{0,S}$ and $F_{\infty,S}$ are $1 - 1$, agreeing on $\partial\Delta$. Our condition also ensures that $M_S(\Gamma)$ is a quasi-Fuchsian group with limit set $F_{0,S}(\partial\Delta)$, [**3**, §5]. The theory of quasi-Fuchsian groups allows that if $p_1, p_2 \in \Delta$ are distinct then fundamental regions for the action of $M_S(\Gamma)$ on $cl(\mathsf{H}^3)$ can be chosen so that the geodesics connecting $F_{\infty,S}(p_j)$ to $F_{0,S}(p_j)$ are in distinct such regions for $j = 1, 2$. But these geodesics are just the images of $\{(z, t) \in cl(\mathsf{H}^3) : z = 0\}$ under action by appropriate elements of $M_S(PSU(1,1)) \subset PSL(2, \mathbb{C})$. Therefore $F_{t,S}(p_1) \neq F_{t,S}(p_2)$ if $p_1 \neq p_2$.

This also ensures that the surfaces are disjoint and that $F_{t,S} = F_{\infty,S}$ on $\partial\Delta$. As $p \to \partial\Delta$ the Euclidean length of the geodesic connecting $F_{0,S}(p)$ to $F_{\infty,S}(p)$ goes to zero, so that

$$\lim_{p \to e^{i\theta}} \|F_{t,S}(p) - F_{\infty,S}(p)\| = 0$$

where length is Euclidean. This concludes the proof.

5. A Comment on the Bers Embedding.

As mentioned preceding Theorem 2, $M_S(\Gamma)$ is a quasi-Fuchsian group with limit set $F_{\infty,S}(\partial\Delta)$. The example from §3 extends to give $F_{\infty,0}(z) = \frac{1}{z}$, $F_{0,0}(z) = \bar{z}$ and $M_0(\Gamma) = \bar{\Gamma}$, as expected.

The Bers embedding $T_{g,n} \hookrightarrow \mathbb{C}^{3g-3+n}$ of Teichmüller space for genus g surfaces with n punctures is constructed by associating to a marked punctured surface $(\mathcal{R}, m)$ a quadratic differential on a fixed marked punctured surface $(\mathcal{R}_0, m_0)$ arising from a marking preserving diffeomorphism $(\mathcal{R}_0, m_0) \to (\mathcal{R}, m)$. On the level of universal covers we obtain an analytic S on Δ satisfying (8) where $\mathcal{R}_0 = \Delta/\Gamma$. It is well known that the forms S, $\|S\|_\infty \leq 1$, are contained in the Bers embedding. See [**3**, §5] for details. We thus have

COROLLARY. *If the conditions of Theorem 2 are met we let* $\tau = \frac{1}{t^2+1}$. *Then the leaves of the foliation give a radial segment in the Bers embedding in the sense that if* $\mathcal{R}_0 = F_{\infty,S}(\Delta)/M_S(\Gamma)$ *and* $\mathcal{R}_1 = \overline{(F_{0,S}(\Delta)/M_S(\Gamma))}$ *then* $\mathcal{R}_\tau = F_{t,S}(\Delta)/M_S(\Gamma)$ *is the surface corresponding to* τS *in the Bers embedding based at* $\mathcal{R}_0$. *(The orientation of* $\mathcal{R}_\tau$ *is as given in the proof of Theorem 1.)*

It is interesting to consider the Ahlfors–Weill mappings if $\|S\|_\infty \geq 1$ but S still occurs in the Bers embedding. Do the sheets $F_{t,S}(\Delta)$ still foliate H^3? What conditions are necessary on a harmonic Beltrami differential so that it occurs via a map $f \colon \mathcal{R}_0 \to \mathcal{R}_1$ homotopic to a diffeomorphism? This is rather unexplored territory and its understanding would lead to more information about the Bers embedding when $\|S\| \geq 1$.

Department of Mathematics, University of Michigan, Ann Arbor, MI 48109-1003

REFERENCES

1. Ahlfors, L. and Weill, G., *A uniqueness theorem for Beltrami equations*, Proc. Amer. Math. Soc. **13** (1962), 975–978.
2. Appell, P., Goursat, E. and Fatou, P., "Théorie des Fonctions Algébriques," Vol. 2, Gauthier–Villars, 1930.
3. Harvey, W.J., "Discrete Groups and Automorphic Functions," Academic Press, 1977.
4. Hille, E., "Ordinary Differential Equations in the Complex Domain," Wiley–Interscience, 1976.
5. Velling, J., *An explicit formula for the monodromy group of a linearly polymorphic function*, in preparation.

Isospectral potentials on
a surface of Genus 3

BY ROBERT BROOKS

Let S be a compact Riemann surface, carrying a metric of constant curvature -1, and let Δ denote the Laplace-Beltrami operator on S.

The question of whether the spectrum of Δ determines the geometry of S is a rather old problem. The first example of surfaces S_1 and S_2 which are isospectral but not isometric was found by Vigneras in [6], by relatively esoteric methods. Shortly thereafter, Sunada [5] found a systematic way of constructing isospectral manifolds, which we will explain briefly below. His method was used by Buser [3] and Brooks-Tse [2] to construct fairly simple examples of isospectral surfaces. In particular, in [2] we showed the following:

THEOREM [2]. *(a) There exist surfaces S_1 and S_2 of genus 3 and variable curvature such that S_1 and S_2 are isospectral but not isometric.*

(b) There exist surfaces S_1 and S_2 of genus 4 and constant curvature which are isospectral but not isometric.

In [4], Guillemin and Kazhdan raised the following related question: given a surface S, are there two distinct smooth functions φ_1 and φ_2 such that the spectrum of $\Delta + \varphi_1$ is equal to the spectrum of $\Delta + \varphi_2$? They showed that if S has simple length spectrum—that is, no two closed geodesics of S have the same length—and negative curvature, then the spectrum of $\Delta + \varphi$ uniquely determines φ. It was subsequently observed that the simple length spectrum condition, while satisfied generically among metrics of variable curvature, was *never* satisfied by a metric of constant negative curvature, so that the theorem of Guillemin and Kazhdan does not apply *at all* to surfaces of constant curvature.

In [1], we constructed the first examples of a surface S with isospectral potentials φ_1 and φ_2. Our method was a variant of that of Sunada.

In the present paper, we will combine the results of [1] and [2] to show:

This work was supported in part by NSF grant DMS-8501300 and an Alfred P. Sloan fellowship.

THEOREM. *There exists a surface S of genus 3 and constant curvature -1, and potentials φ_1 and φ_2 such that $\Delta + \varphi_1$ is isospectral to $\Delta + \varphi_2$.*

The surface we construct is quite simple—it is a 14-fold ramified covering of the $(7,7,7)$-triangle group. The potentials are also quite easy to describe. We hope that the simplicity of our construction, as well as its explicit nature, will shed light on the general problem of isospectrality.

We remark that the following natural problem is still open, as of this writing: find a Riemann surface of hyperbolic type for which the spectrum of $\Delta + \varphi$ uniquely determines φ, for all potentials φ. One expects that a generic Riemann surface should have this property.

In order to prove the theorem, we recall briefly the previous work of Sunada [5] and Brooks-Tse [1], [2]. Let us say that a triple of finite groups (G, H_1, H_2) satisfies the conjugacy condition if:

(i) H_1 and H_2 are non-conjugate subgroups of G
(ii) for all elements $g \in G$,

$$\sharp([g] \cap H_1) = \sharp([g] \cap H_2)$$

where $[g]$ denotes the conjugacy class of g in G.

Suppose we are given a manifold M and a surjective homomorphism $f : \pi_1(M) \to G$. Let M^{H_1}, M^{H_2}, and M^G denote the Riemannian coverings of M whose fundamental groups are $f^{-1}(H_1)$, $f^{-1}(H_2)$, and $f^{-1}(id)$ respectively.

Then Sunada proved the following:

THEOREM [5]. *Under these conditions, M^{H_1} and M^{H_2} are isospectral.*

In general, of course, M^{H_1} and M^{H_2} may happen to be isometric as well, but by arguing separately in the various cases he considered, Sunada showed that that does not happen very often.

In [1], we showed that, in the unlikely event that M^{H_1} and M^{H_2} are isometric, one may still construct isospectral potentials. The thrust of our argument was:

(a) Sunada's argument extends easily when one lifts, not only Δ, but also the operator $\Delta + \varphi$ from M to M^{H_1} and M^{H_2}.
(b) For generic choices of φ, the isometry from M^{H_1} to M^{H_2} will not preserve the lift of φ from M to M^{H_1} and M^{H_2}.

We also gave explicit examples of finite groups for which one could conclude that M^{H1} is isometric to M^{H2}, thus giving examples of isospectral potentials.

In [2], we considered the following construction: let G be the group $SL(3, Z/2)$, and H_1, H_2 the subgroups

$$H_1 = \begin{pmatrix} * & * & * \\ 0 & * & * \\ 0 & * & * \end{pmatrix} \qquad H_2 = \begin{pmatrix} * & 0 & 0 \\ * & * & * \\ * & * & * \end{pmatrix}$$

Then H_1 and H_2 satisfy the conjugacy condition.

If we now denote by Γ the subgroup of index 2 of the $(7, 7, 7)$ triangle group consisting of orientation-preserving isometries of H^2, then Γ is given by generators and relations by $\Gamma = \{A, B : A^7 = B^7 = (AB)^7 = id\}$.

There is then a surjective homomorphism $\Gamma \to G$ given by

$$A \to \begin{pmatrix} 1 & 0 & 1 \\ 1 & 1 & 0 \\ 0 & 1 & 0 \end{pmatrix} \qquad B \to \begin{pmatrix} 0 & 0 & 1 \\ 1 & 0 & 0 \\ 1 & 1 & 0 \end{pmatrix}$$

$$AB \to \begin{pmatrix} 1 & 1 & 1 \\ 1 & 0 & 0 \\ 1 & 1 & 0 \end{pmatrix}.$$

We then observed that Sunada's argument worked perfectly well when M is an orbifold, and $\pi_1(M)$ its orbifold fundamental group, provided only that M^{H1} and M^{H2} are smooth manifolds.

When one carries out this argument for H^2/Γ, one checks easily that M^{H1} and M^{H2} are smooth surfaces of genus 3. In Figures 1 and 2 below, taken from [2], we give gluing diagrams for M^{H1} and M^{H2}.

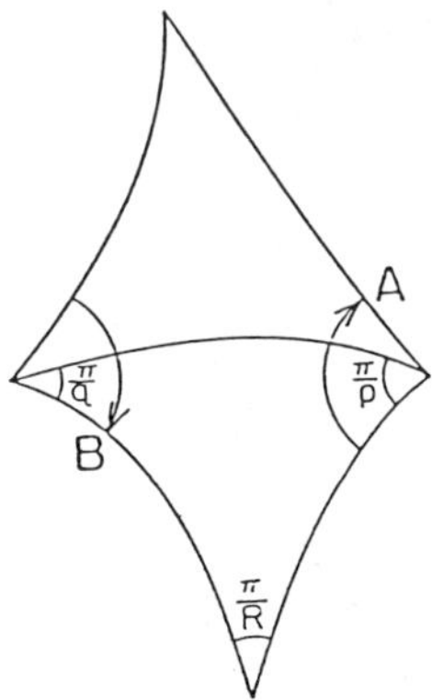

Figure 1: $p = q = r = 7$

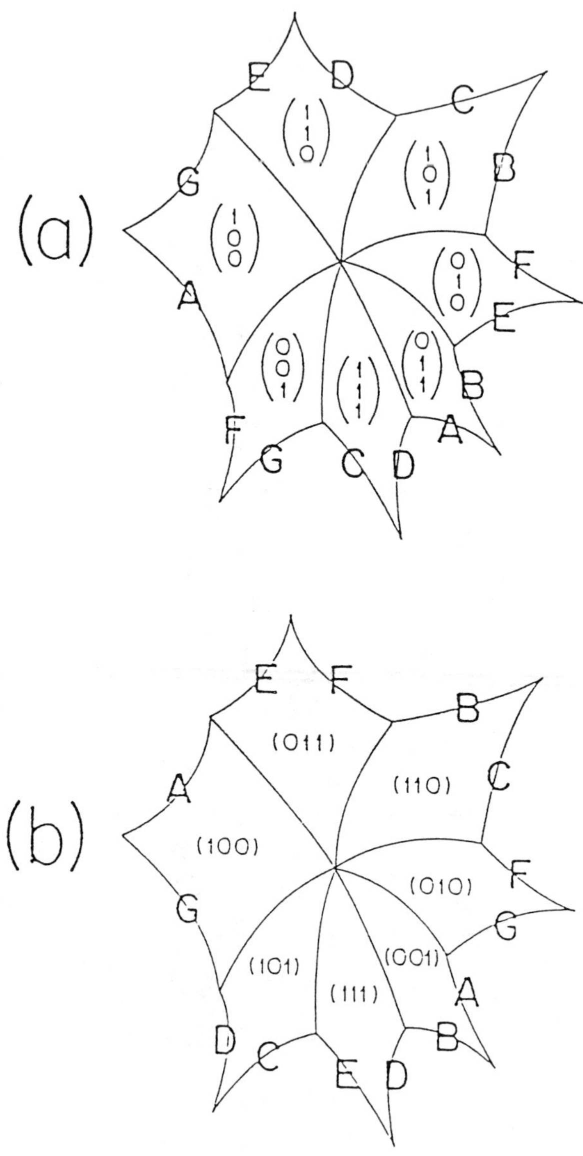

Figure 2

It turns out that if H^2/Γ is given its unique orbifold metric of constant curvature -1, then M^{H1} and M^{H2} are isometric by an orientation-reversing isometry. This can be verified by changing the labels on 2(a) according to the following scheme:

$$A \to B$$
$$B \to D$$
$$C \to G$$
$$D \to A$$
$$E \to E$$
$$F \to C$$
$$G \to F$$

When one then holds 2(a) up to a mirror, one obtains 2(b).

One now chooses a function φ on H^2/Γ which is not invariant under orientation-reversing reflection. When φ is lifted to $M^{H1} \cong M^{H2}$, one obtains the desired potentials φ_1 and φ_2 such that $\Delta + \varphi_1$ is isospectral to $\Delta + \varphi_2$ on M^{H1}.

This completes the proof of the theorem.

Department of Mathematics, University of Southern California,
Los Angeles, CA 90089-1113

REFERENCES

1. Brooks, R., *On Manifolds of Negative Curvature with Isospectral Potentials*, Topology **26** (1987), 63–66.
2. Brooks, R., and Tse, R., *Isospectral Surfaces of Small Genus*, Nagoya Math J. **107** (1987), 13–24.
3. Buser, P., *Isospectral Riemann Surfaces*, Ann. Inst. Fourier XXXVI (1986), 167–192.
4. Guillemin, V., and Kazhdan, D., *Some Inverse Spectral Results for Negatively Curved 2-Manifolds*, Topology **19** (1980), 301–312.
5. Sunada, T., *Riemannian Coverings and Isospectral Manifolds*, Ann. Math **121** (1985), 169–186.
6. Vigneras, M.F., *Varietes Riemanniennes Isospectrales et non Isometriques*, Ann. Math **112** (1980), 21–32.

Integrable holomorphic quadratic differentials
with simple zeros

BY YUKIO KUSUNOKI

On every compact Riemann surface of genus greater than one, there exist holomorphic differentials and holomorphic quadratic differentials having only simple zeros. The usefulness of such differentials in Teichmüller Theory is well known (cf. L. Ahlfors [1], L. Bers [2]).

The purpose of this short note is to give some analytic version for the existence of such differentials on an arbitrary (compact or noncompact) Riemann surface. In particular, our result will give an analytic proof for the above classical result from algebraic geometry.

The author is grateful to M. Taniguchi and the referee for valuable comments.

1. Notations and Results. Let X be an arbitrary Riemann surface of genus $g(\leq \infty)$ and $W^1(X)(= \Gamma_a(X))$ be the Hilbert space of square integrable holomorphic abelian differentials on X, and $W^2(X)$ be the Banach space of integrable holomorphic quadratic differentials on X. If X is compact, $W^1(X)(W^2(X))$ is just the complex vector space of holomorphic abelian (quadratic) differentials on X. It is known that if $g = \infty$, then $d_1 = \dim W^1(X)$ is infinite, while if $g < \infty$, then $d_1 = \infty$ or $d_1 = g$, where the latter case happens only if X is compact or X is of parabolic type (i.e. $X \in O_G$) and g is finite. Let $d_2 = \dim W^2(X)$. For Riemann surface X with $d_1 = \infty$ we have $d_2 = \infty$. Thus d_2 is finite only if X is a compact Riemann surface with a finite number of punctures. Actually then $d_2 = 3g - 3 + n$ for n-punctured (compact) Riemann surface of genus $g > 1$.

Both $W^1(X)$ and $W^2(X)$ are nontrivial if $g \geq 1$. The exceptional case for which they become trivial occurs only for X with $g = 0$, more precisely, $W^1(X)$ is trivial for $X \in O_G$ with $g = 0$, and so $W^2(X)$ is for X conformally equivalent to $\hat{C}$ with at most three punctures. In the following we shall exclude these exceptional cases.

The main result is as follows.

THEOREM. *Let X be an arbitrary Riemann surface of genus $g(0 \leq g \leq \infty)$. Then for each $j(= 1, 2)$, the set of all differentials of $W^j(X)$ with only simple zeros on X is dense in $W^j(X)$.*

Here by a differential ω with only simple zeros on X we mean that ω has simple zeros whenever it has zeros on X. Contrary to the case of compact surfaces, $W^j(X)$ for noncompact X may contain the differentials with an infinite number of zeros or without zeros.

COROLLARY. *If X is compact, then the set of holomorphic abelian (quadratic) differentials with $2g - 2(4g - 4)$ simple zeros is dense in $W^1(X)(W^2(X))$.*

REMARK: From the theorem every $\omega \in W^j(X)$ can be approximated locally uniformly on X by $\omega_n \in W^j(X)$ with only simple zeros. For noncompact X, let ν and ν_n be the numbers of zeros (counted with multiplicities) of ω and ω_n. So far as these numbers we just know that $\nu \leq \nu_n \leq \infty$ for large n if $\nu < \infty$, and $\{\nu_n\}$ tend to ∞ if $\nu = \infty$. (As for the growth of ν, see e.g. [**3**]).

2. For the proof of theorem we provide the following lemmas. In the sequel X denotes an arbitrary Riemann surface and we write simply W^j for $W^j(X)$.

LEMMA 1. *For each $j(= 1, 2)$, the differentials of W^j have no common zeros on X, provided that W^j is nontrivial.*

PROOF: If X is compact, the assertion is an immediate consequence from Riemann–Roch theorem. Now consider the nontrivial W^1. The following proof is valid for arbitrary X (cf. K.I. Virtanen [**5**]). Suppose there is a common zero $q \in X$ for all differentials of W^1, then the Bergman kernel differential $K(z, \varsigma)dz$ for the space W^1 vanishes at $q(\leftrightarrow \varsigma)$ and $\|K(z, \varsigma)dz\|^2 = K(\varsigma, \varsigma) = 0$, hence $K(z, \varsigma)dz = 0$. On the other hand we know

$$K(z, \varsigma)dz = -\frac{1}{2\pi}(\psi_q^1 + i\tilde{\psi}_q^1)$$

where $\psi_q^1 = 2(\partial u/\partial z)dz$, $\tilde{\psi}_q^1 = 2(\partial \tilde{u}/\partial z)dz$ and $u(\tilde{u})$ is the normalized potential on X which is harmonic except a singularity $\mathrm{Re}(1/(z - \varsigma))(\mathrm{Im}(-1/(z - \varsigma))$ at q respectively (cf. R. Nevanlinna [**4**]). Thus the integral $f(= u - i\tilde{u})$ of $\psi_q^1 = -i\tilde{\psi}_q^1$ is single-valued analytic on X $\{q\}$, has a simple pole at q and bounded outside of a neighborhood of q. Take any $\omega(\not\equiv 0) \in W^1$, then ω has a zero of order $k(\geq 1)$ at q and $f^k\omega \in W^1$ does not vanish at q, which is a contradiction.

As for nontrivial W^2, first suppose W^1 is also nontrivial. For any point $q \in X$ take a differential $\sigma \in W^1$ which does not vanish at q. Then $\sigma^2 \in W^2$

does not vanish at q. It remains the case $X \in O_G$ with $g = 0$, but W^2 is not trivial. We may assume that X is a domain in $\hat{C}$ whose boundary contains four points z_1, z_2, z_3 and ∞. Then $\prod_{i=1}^{3}(z - z_i)^{-1}dz^2 \in W^2$ does not vanish at any point of X. $\qquad\square$

LEMMA 2. *For a point q in X, let $W^j(q)(= W^j(q, X))$ be the subspace of W^j whose elements have a zero at q, then $W^j(q)$ are nowhere dense in W^j.*

PROOF: It suffices to show that $W^j(q)$ does not have an interior point in W^j. Let θ be an element of $W^j(q)$, then there is an $\omega_q \in W^j$ which does not vanish at q (Lemma 1). Putting $\sigma = \lambda\omega_q + (1 - \lambda)\theta(0 < \lambda < 1)$, then $\sigma \in W^j\backslash W^j(q)$ and $\|\theta - \sigma\| \leq \lambda(\|\omega_q\| + \|\theta\|)$ in respective norm of W^j. Hence σ is in a neighborhood of θ provided λ is sufficiently small. $\qquad\square$

LEMMA 3. *Let E be a discrete (countable) set of points on X. Then there exist the differentials of W^1 and W^2 which do not vanish at all points of E.*

PROOF: Let $E = \{q_k\}$. For each k, $W^j(q_k)$ are nowhere dense in $W^j(j = 1, 2)$. While, a complete metric space is of the second Baire category. It follows that

$$W^j \neq \bigcup_k W^j(q_k)$$

by Baire's category theorem (cf. [6]), which completes our assertion. $\qquad\square$

REMARK: Let $W^j(\delta)$ be the subspace of W^j whose elements are the multiple of an integral divisor δ. Then by Lemma 3, Lemma 2 is also valid for $W^j(\delta)$ instead of $W^j(q)$.

3. Proof of Theorem. Let θ be any fixed differential of W^j. It suffices to consider θ with zeros. By Lemma 3 we take a differential $\sigma \in W^j$ which does not vanish at all zeros of θ. Then a meromorphic function

$$f = \frac{\theta}{\sigma}$$

gives the projection map of X into $\hat{C}$. Given $\varepsilon > 0$, we choose a sufficiently small complex number λ such that $0 < |\lambda| < \varepsilon/\|\sigma\|$ and $f^{-1}(\lambda)$ does not contain any branch points. (If the number (with multiplicities) of zeros of θ is $\geq n$, then $f^{-1}(\lambda)$ for small λ contains at least n points in the neighborhoods of zeros of θ). Now

$$\omega = \theta - \lambda\sigma$$

belongs to W^j and $\|\theta - \omega\| < \varepsilon$. It remains to prove that ω has a simple zero at every possible zeros of ω. From the choice of σ one sees that if q is a zero of ω, then both θ and σ do not have a zero at q, and $f(q) = \lambda$. It follows that in terms of local parameter z at q, $\sigma = \sigma(z)dz^j = (b_0 + b_1 z + \ldots)dz^j$ with $b_0 \neq 0$, $f(z) = \lambda + c_1 z + \ldots$ with $c_1 \neq 0$ and thus

$$\omega = (f - \lambda)\sigma = (b_0 c_1 z + \ldots)dz^j$$

has a simple zero at q, which completes the proof. $\qquad\square$

Yukio Kusunoki, Department of Mathematics, Faculty of Science, Kyoto University

REFERENCES

1. Ahlfors, L., The complex analytic structure of the space of closed Riemann surfaces, in "Analytic functions", pp. 45–66, Princeton Univ. Press, Princeton, 1960.
2. Bers, L., *Holomorphic differentials as functions of moduli*, Bull. Amer. Math. Soc. **67** (1960), 206–210.
3. Kusunoki, Y., *Beiträge zur Theorie der analytischen Differentiale und Funktionen*, Kōdai Math. Sem. Rep. **26** (1975), 446–453.
4. Nevanlinna, R., "Uniformisierung," Springer Verlag, Berlin Heidelberg New York, 2nd ed., 1967.
5. Virtanen, K.I., *Über Extremalfunktionen auf offenen Riemannschen Flächen*, Ann. Acad. Sci. Fenn. Ser. A.I., No. 141 (1952) 7pp.
6. Yosida, K., "Functional analysis," Springer Verlag, 2nd ed., 1968.

Lower bounds for the number of saddle connections and closed trajectories of a quadratic differential

BY HOWARD MASUR

Dedicated to the memory of my father, Ernest F. Masur.

§. Introduction.

Suppose X is the torus the complex plane C divided by the group of translations generated by $z \to z + 1$ and $z \to z + i$ and $q = dz^2$ is the unique up to scalar multiple quadratic differential on X. The trajectories of q are straight lines and a trajectory is closed if and only if its slope is a rational p/q. Its length is $(p^2 + q^2)^{1/2}$. Parallel closed trajectories fill up X. The number $N(T)$ of parallel families of length $\leq T$ is then the number of lattice points (p, q) with p, q relatively prime inside a circle of radius T. It is classical that

$$\lim_{T \to \infty} \frac{N(T)}{T^2} = \frac{6}{\pi^2}.$$

Now suppose X is a closed surface of genus ≥ 2 and $q_0(z)dz^2$ is a nonzero holomorphic quadratic differential on X. Now q_0 has zeroes. We are interested in two asymptotic counts. Let $N_1(T)$ denote the number of singular trajectories of length $\leq T$ joining two zeroes of q_0 and let $N_2(T)$ denote the number of parallel families of closed regular trajectories of length $\leq T$.

THEOREM 1. $0 < \underline{\lim_{T \to \infty}} \dfrac{N_1(T)}{T^2} \leq \overline{\lim}_{T \to \infty} \dfrac{N_1(T)}{T^2} < \infty.$

THEOREM 2. *There exists $c > 0$ such that*

$$c\, N_1(T) \leq N_2(T) \leq N_1(T).$$

Corollary. Suppose Δ is a rational billiard table (see [K-M-S] for definition). The number of generalized diagonals and parallel families of closed orbits grow quadratically in the sense of Theorems 1 and 2.

The upper bound in Theorem 1 was proved in [M2]. The upper bound in Theorem 2 is trivial since for every annulus of parallel closed trajectories there is a singular trajectory on the boundary of the annulus of shorter

---

This work was supported in part by NSF Grant DMS-8601977.

length. The fact that $N_2(T) \neq 0$ for some T was first proved in [**M1**]. In this paper we will prove the lower bounds in each theorem.

The quadratic growth rate is perhaps surprising since the number of geodesics of length $\leq T$ corresponding to free homotopy classes of simple closed curves grows like T^{6g-6} [**R**]. The point here is "most" such geodesics are made up of pieces of singular trajectories; the number of singular trajectories is much less. The point of Theorem 2 is a fixed percentage of singular trajectories lie on the boundary of annuli of closed trajectories that are not much longer.

I don't know if there are actually limits in the growth rates as there are for a torus.

Section 1 is devoted to preliminaries of quadratic differentials, the metric and interval exchange transformation they define; §2 to the proof of the lower bound in Theorem 1. It depends on the upper bound. I am indebted to Michal Boshernitzan for pointing out to me that $N_1(T)$ grows at least quadratically for subsequences of T and the condition on subsequences can be removed once an upper bound is known.

§3 recalls certain techniques and results from [**M1**] and [**M2**] needed here. §4 gives the proof of the lower bound in Theorem 2.

I would like to thank the referee for urging me to make substantial revisions in the original manuscript.

§1. Preliminaries.

We let X be a closed Riemann surface of genus $g = 2$ and $q_0 = q_0(z)dz^2$ a nonzero holomorphic quadratic differential on X such that $\int_X |q_0| = 1$. It defines a metric $|q_0^{1/2}dz|$ on X. A trajectory is an arc along which $\arg q_0(z)dz^2$ is constant. Horizontal trajectories are defined by $\arg = 0$, vertical trajectories by $\arg = \pi$. A trajectory is regular if it does not contain zeroes of q_0 and singular if the two endpoints of the trajectory are zeroes and there are no zeroes in the interior of the arc. Singular trajectories we will also call saddle connections. If there is a closed regular trajectory there is a parallel family of freely homotopic closed trajectories of equal length sweeping out an annulus. On the boundary of the annulus are saddle connections of smaller length.

At a zero of order $k \geq 0$ of q_0 there are local coordinates so that $q_0(z)dz^2 = z^k dz^2$. If $k = 0$ the metric is locally Euclidean. A zero of order $k > 0$ has a neighborhood which is isometric to a neighborhood of the origin in $\mathbf{R}^2$ with the metric $ds^2 = dr^2 + ((1 + k/2)rd\theta)^2$. The total angle at the zero is $\pi(2+k)$ and a geodesic through the zero consists of two lines which make an angle of at least π at the zero. In each free homotopy class of a simple curve there is either a unique geodesic or an annulus of closed regular trajectories.

Now for each $\theta \in [0, 2\pi)$ choose a finite set of horizontal trajectories ℓ_θ for $e^{i\theta} q_0$ with the properties that

i) one endpoint of each ℓ_θ is a zero of q_0.

ii) The other endpoint is either a zero or the vertical trajectory through it in one of the two directions hits a zero before returning to any ℓ_θ.

iii) Every vertical trajectory that is not a saddle connection if sufficiently extended intersects some ℓ_θ.

The ℓ_θ define a decomposition of X (cf. [S]) into rectangles R_j of width h_j and height v_j each swept out by vertical trajectories with endpoint on ℓ_θ. The vertices of R_j on ℓ_θ are either endpoints of ℓ_θ, zeroes of q_0 or points such that the vertical trajectory through the point in one direction hits a zero before returning to ℓ_θ. Let V denote the set of such vertices. The number of vertices hence the number of rectangles is bounded in terms of the genus.

Label the two sides of each ℓ_θ as ℓ_θ^+ and ℓ_θ^- and points as p^+ and p^-. The decomposition of X into rectangles defines a map

$$T : \bigcup_{\ell_\theta} (\ell_\theta^+ \cup \ell_\theta^-) - V \to \bigcup_{\ell_\theta} (\ell_\theta^+ \cup \ell_\theta^-) - V.$$

For $p^+ \in \ell_\theta^+$ if the vertical trajectory through p^+ next returns to some ℓ_θ at q^- (resp. q^+) define $T(p^+) = q^+$ (resp q^-). Similarly define T at $p^- \in \ell_\theta^-$. T is a local isometry called a (generalized) interval exchange.

For θ' near θ there are corresponding $\ell_{\theta'}$ with the same properties. For each rectangle of $e^{i\theta} q_0$ there is a nearby rectangle of $e^{i\theta'} q_0$. If $e^{i\theta} q_0$ has vertical saddle connections not intersecting ℓ_θ, $e^{i\theta'} q_0$ will define additional rectangles with small width and heights bounded in terms of the heights of the rectangles of $e^{i\theta} q_0$. Thus the heights are locally bounded. By compactness of $[0, 2\pi]$ we may assume there are upper bounds to the widths and heights of the rectangles.

Now let θ be such that $e^{i\theta} q_0$ has dense vertical trajectories. This will be true for all but countably many θ. Consider the iterated map $T^{(n)}$ defined on $\bigcup(\ell_\theta^+ \cup \ell_\theta^-) - \bigcup\limits_{j=0}^{n-1} T^{-j}(V)$. The points of the latter union are all distinct since $e^{i\theta}q_0$ has no vertical saddle connections. $T^{(n)}$ is again an interval exchange. The number of rectangles R^n defined by $T^{(n)}$ grows linearly with n. This means there is a constant M_1 and a rectangle R^n with width at most M_1/n. Under iteration the points of R^n all visit the jth rectangle of the original decomposition the same number of times, say $n_j \leq n$. If its height is v_j, the height of R^n is then

$$\sum n_j v_j \leq nM_2 \text{ for some constant } M_2.$$

On the vertical sides of R^n are zeroes z_1, z_2 of q_0 unless one of the two horizontal sides of R^n contains an endpoint of ℓ_θ in which case a slightly larger rectangle has zeroes on both vertical sides. Let Z be the saddle connection crossing R^n joining z_1, z_2 and let $|Z|$ denote its length. Since the length is less than the sum of the width and height of R^n

$$|Z| \leq nM_2 + M_1/n.$$

Let θ' be the angle Z makes with the vertical direction. Then

$$|\sin \theta'| \leq \frac{M_1/n}{|Z|}.$$

There is a lower bound ϵ to the length of any saddle connection. We assume n sufficiently large so

$$\theta' \leq \frac{2M_1/n}{|Z|} \text{ and } |Z| \leq 2M_2 n.$$

Thus in the $\frac{2M_1}{n|Z|}$ neighborhood of every such θ there is a saddle connection of length $\leq 2M_2 n$.

For any θ' such that $e^{i\theta'}q_0$ has a vertical saddle connection Z of length $\leq 2M_2 n$ consider the interval of width $\frac{2M_1}{n|Z|}$ about θ'. We have shown these intervals cover $[0, 2\pi]$ up to measure 0. We have shown:

PROPOSITION 1.1. *There is a constant* $c_1 > 0$ *such that for all* n,

$$\sum_{|Z| \leq n} \frac{1}{|Z|} \geq c_1 n.$$

§2. Proof of Theorem 1 (lower bound).

We know from [**M2**] there is a constant c_2 such that

$$(2.1) \qquad N_2(T) \leq N_1(T) \leq c_2 T^2.$$

There is a lower bound ϵ to the length of any saddle connection on X. Fix $0 < \sigma < 1$ and for any $n > 0$, $k > 0$ break the interval $[\epsilon, \sigma^k n)$ into a union of intervals

$$[\epsilon, \sigma^{j_0} n), \ [\sigma^{j_0} n, \ \sigma^{j_0 - 1} n), \dots, [\sigma^{k+1} n, \ \sigma^k n)$$

for some uniquely defined j_0. Let N_j be the number of saddle connections in the interval with right endpoint $\sigma^j n$. From 2.1 we have

$$N_j \leq c_2 \sigma^{2j} n^2$$

Then

$$
\begin{aligned}
(2.2) \qquad \sum_{\epsilon \leq |Z| \leq \sigma^k n} \frac{1}{|Z|} &\leq \sum_{k \leq j \leq j_0} \frac{1}{\sigma^{j+1} n} N_j \\
&\leq \sum_{k \leq j \leq j_0} \frac{c_2 \sigma^{2j} n^2}{\sigma^{j+1} n} \\
&\leq c_2 \sum_{j \geq k} \sigma^{j-1} n \\
&= c_2 n \frac{\sigma^{k-1}}{1 - \sigma}.
\end{aligned}
$$

Now suppose $\displaystyle \lim_{T \to \infty} \frac{N_1(T)}{T^2} = 0$. Choose a sequence n_k for which $\displaystyle \lim \frac{N_1(n_k)}{n_k^2} = 0$. Choose k so that $c_2 \dfrac{\sigma^{k-1}}{1 - \sigma} \leq \dfrac{c_1}{3}$ where c_1 is given in Proposition 1.1. Choose $\delta > 0$ so $\delta/\sigma^k \leq c_1/3$ and n_j so that $\dfrac{N_1(n_j)}{n_j^2} < \delta$.

Then

$$
\begin{aligned}
\sum_{|Z| \leq n_j} \frac{1}{|Z|} &= \sum_{|Z| \leq \sigma^k n_j} \frac{1}{|Z|} + \sum_{\sigma^k n_j < |Z| \leq n_j} \frac{1}{|Z|} \\
&\leq \frac{c_1}{3} n_j + \frac{1}{\sigma^k n_j} N_1(n_j) \quad \text{by (2.2) and the choice of } k \\
&\leq \frac{c_1}{3} n_j + \frac{1}{\sigma^k n_j} \delta n_j^2 \leq \frac{2c_1}{3} n_j \quad \text{which contradicts Proposition 1.1.}
\end{aligned}
$$

COROLLARY 2.2. *There are constants $c_3 > 0$ and $0 < \sigma_0 < 1$ such that for T sufficiently large*

$$N_1(T) - N_1(\sigma_0 T) \geq c_3 T^2.$$

PROOF: $N_1(\sigma_0 T) < c_2 \sigma_0^2 T^2$ and $N_1(T) \geq cT^2$ for some $c > 0$ and T sufficiently large.

§3. Closed Trajectories and surfaces with boundary.

In this section we will allow Y to be a finite Riemann surface, possibly disconnected, of genus g with n punctures; φ a nonzero integrable meromorphic quadratic differential on Y with area 1. In particular φ has at most simple poles at the punctures. A saddle connection will now include any trajectory to a puncture even from a nonzero point. A collection of saddle connections is *lined up* if all members are defined by the same argument.

PROPOSITION 3.1. *Given a lined up collection of saddle connections on Y there is a closed trajectory α disjoint from every saddle connection.*

PROOF: Consider the angle θ so that each saddle connection is a vertical trajectory of $e^{i\theta}\varphi$ and then consider the Teichmüller map $f_K : Y \to Y_K$ with dilatation

$$\frac{K-1}{K+1} \quad \overline{e^{i\theta}\varphi}/|\varphi|$$

and terminal differential of unit norm φ_K on Y_K. We can assume by taking subsequences that in an appropriate compactified moduli space that

$$Y_K \to Y_\infty, \varphi \to \varphi_\infty \text{ as } K \to \infty$$

where φ_∞ is a quadratic differential on Y_∞.

If $\varphi_\infty \equiv 0$ then it was shown in Lemma 2.1 of [M1] that φ_K has a closed trajectory for K large and the width across the annulus of closed trajectories goes to infinity with K. Since the length of each saddle connection goes to zero as $K \to \infty$, no saddle connection can cross the annulus so the closed trajectories are disjoint. This is then true of $e^{i\theta}\varphi$ as well since we simply map back by the Teichmüller map f_K^{-1}.

If $\varphi_\infty \not\equiv 0$ there is a component of Y_∞ with a nonzero differential. It has a family of closed trajectories by Theorem 2 of [M1]. Then φ_K does as well for large K and no saddle connection in the family can cross the annulus of closed trajectories since the width of the annulus is bounded below. □

Now suppose β is a saddle connection of unit length on Y, a closed surface. Let $\alpha = \alpha(\beta)$ denote the shortest closed trajectory on Y disjoint from β, which exists by Proposition 3.1. Denote by $|\alpha|_\varphi$ the φ length of α.

LEMMA 3.2. $\displaystyle\sup_{||\varphi||=1}\;\sup_{|\beta|_\varphi=1}\;|\alpha|_\varphi = M < \infty.$

The lemma says that as φ and β vary over the set of unit norm quadratic differentials and unit length saddle connections, the shortest closed trajectory has uniformly bounded length.

PROOF: Suppose $|\alpha_n(\beta_n)|_{\varphi_n} \to \infty$ for a sequence of φ_n on Y_n and saddle connections β_n. Again we may assume $Y_n \to Y_\infty$, $\varphi_n \to \varphi_\infty$.

If Y_∞ is compact a subsequence of β_n converges to a unit length saddle connection β_∞ on Y_∞. There is a closed trajectory α_∞ on Y_∞ disjoint from β_∞ by Proposition 3.1. Then there is for large n a closed trajectory on Y_n of bounded length disjoint from β_n, a contradiction.

Thus assume Y_∞ has been pinched along curves $\gamma_1, \ldots \gamma_p$. If $\varphi_\infty \equiv 0$ then by Lemma 2.1 of [M1], φ_n has closed trajectories of length $\to 0$ in the homotopy class of some γ_i and the width of the annulus $\to \infty$. The β_n cannot cross this annulus and again we contradict the shortest α_n going to infinity in length.

Finally if $\varphi_\infty \not\equiv 0$ a subsequence of β_n converges to a lined up collection of saddle connections on a union of components on which $\varphi_\infty \not\equiv 0$ or there is a component of Y_∞ disjoint from β_n on which $\varphi_\infty \not\equiv 0$. In the first case, by Proposition 3.1 there is a closed trajectory α_∞ disjoint from the collection, hence a closed trajectory α_n disjoint from β_n of bounded length. In the second case there is a closed trajectory α_∞ on the component of Y_∞ and therefore for large n, a closed trajectory α_n disjoint from β_n, again of bounded length, a contradiction. $\square$

Next we adopt the following notation. We have fixed $0 < \sigma_0 < 1$ given in Corollary 2.2. $S(T)$ denotes the set of saddle connections with respect to q_0 on X with lengths in $[\sigma_0 T, T]$. $C(T)$ denotes the set of parallel families of closed trajectories with respect to q_0 with length $\leq T$. $S(\varphi, T)$ denotes the set of saddle connections of length T with respect to a varying metric φ. $C(\varphi, T)$ is the set of families of closed trajectories with length $\leq T$.

Now Lemma 3.2 allows us to define a map

$$h : S(\varphi, 1) \to C(\varphi, M)$$

associating to any β of unit length, a closed trajectory of length $\leq M$. Now let $\beta \in S(T)$. Let θ_β be the angle so β is vertical for $e^{i\theta_\beta} q_0$. Contract β via the Teichmüller map determined by $e^{i\theta_\beta} q_0$ so β has unit length on the terminal surface Y_β with respect to the terminal differential φ_β. Denote by

$|\ |_\beta = |\ |_{\varphi_\beta}$ lengths with respect to this metric. The association of β on X to β on Y_β gives a map

$$H : S(T) \to \cup_\varphi S(\varphi, 1).$$

We will speak of β as both a saddle connection on X and on Y_β. The heart of the proof of Theorem 2 in the next section is the computation of $\mathrm{card}(h \circ H)^{-1}(\alpha)$ for each $\alpha \in C(\varphi, M)$.

We first need the following concept and result from [**M2**].

Definition. A system $\Gamma = (\gamma, \dots, \gamma_p)$ of saddle connections L separates β_1 from β_2 if

  i) γ_i, γ_j are disjoint except possibly at zeroes, $i \neq j$.

  ii) Γ is disjoint from β_i except possibly at zeroes.

  iii) Any arc from β_1 to β_2 intersects Γ.

  iv) $|\beta_2|/|\Gamma| \geq L$ where $|\Gamma| = \max\limits_{\gamma_i \in \Gamma} |\gamma_i|$.

Note Γ may contain β_1. An important and easy fact is there is a number C_1 depending only on the genus such that if $\Gamma = (\gamma_1, \dots, \gamma_p)$, $p > C_1$ is a system of saddle connections, two of the γ must cross at a nonzero point.

LEMMA 3.3. *([M2, Proposition 1.3]). Suppose Γ_1 and Γ_2 are two separating systems. Then there is a separating system Γ, called the combination of Γ_1 and Γ_2 with the properties that*

  i) $|\Gamma| \leq C_1(|\Gamma_1| + |\Gamma_2|)$.

  ii) *If β is disjoint from each Γ_i except for zeroes, then Γ separates β from Γ_i.*

Note this implies that if Γ_1 say, separates β and α then Γ does as well.

§4. The Proof of Theorem 2 (lower bound).

We begin with the following idea. Suppose $\beta_1, \beta_2 \in (h \circ H)^{-1}(\alpha)$. By definition $|\beta_i|_{\beta_i} = 1$. However it is possible $|\beta_2|_{\beta_1} \geq 1$. Therefore for any subset $S' \subset (h \circ H)^{-1}(\alpha)$ let

$$M_\alpha = \sup_{\beta \in S'} |\alpha|_\beta \leq M \text{ and let } \beta' \in S'$$

satisfy $|\alpha|_{\beta'} \geq M_\alpha/2$. Again denoting by $|\ |$ lengths on X, the base surface, we have:

LEMMA 4.1. *For any $\beta \in S'$ and T sufficiently large,*

$$\frac{|\beta|_{\beta'}}{|\alpha|_{\beta'}} \leq \frac{8T}{\sigma_0 |\alpha|}.$$

PROOF: Let $\theta_{\alpha,\beta} = |\theta_\alpha - \theta_\beta|$, the angle between the vertical angles θ_α and θ_β. Let $h_{\alpha,\beta} = \int_\alpha |\mathrm{Re}(e^{i\theta_\beta} q_0)^{1/2}|$ and $v_{\alpha,\beta} = \int_\alpha |\mathrm{Im}(e^{i\theta_\beta} q_0)^{1/2}|$ the horizontal and vertical lengths of α with respect to $e^{i\theta_\beta} q_0$. After the Teichmüller map with maximal dilatation K satisfying $\sigma_0 T \leq K^{1/2} \leq T$ we have $|\alpha|_\beta \leq M_\alpha$. This implies $K^{1/2} h_{\alpha,\beta} \leq M_\alpha$ which gives

$$\sin \theta_{\alpha,\beta} = \frac{h_{\alpha,\beta}}{|\alpha|} \leq \frac{M_\alpha}{\sigma_0 T |\alpha|}.$$

Therefore

$$\frac{h_{\beta,\beta'}}{|\beta|} = \sin \theta_{\beta,\beta'} \leq \sin \theta_{\beta,\alpha} + \sin \theta_{\beta',\alpha} \leq \frac{2M_\alpha}{T\sigma_0 |\alpha|}$$

and

$$K^{1/2} h_{\beta,\beta'} \leq T h_{\beta,\beta'} \leq \frac{2|\beta| M_\alpha}{\sigma_0 |\alpha|} \leq \frac{2T M_\alpha}{\sigma_0 |\alpha|} \leq \frac{4T |\alpha|_{\beta'}}{\sigma_0 |\alpha|}.$$

On the other hand

$$v_{\beta,\beta'} \leq |\beta| \leq T$$

so

$$\frac{v_{\beta,\beta'}}{K^{1/2}} \leq \frac{v_{\beta,\beta'}}{\sigma_0 T} \leq \frac{T}{\sigma_0 T} = \frac{1}{\sigma_0} \leq \frac{4T |\alpha|_{\beta'}}{\sigma_0 |\alpha|}$$

for T sufficiently large. Thus

$$|\beta|_{\beta'} \leq \frac{v_{\beta,\beta'}}{K^{1/2}} + K^{1/2} h_{\beta,\beta'} \leq \frac{8T |\alpha|_{\beta'}}{\sigma_0 |\alpha|}$$

for sufficiently T large. $\square$

Now Theorem 4.1 of [**M2**] allows us to compute for a given saddle connection or closed trajectory α the number of saddle connections β disjoint from α of a given length. We would like to do this for the surface $Y_{\beta'}$ and metric $|\ |_{\beta'}$ using Lemma 4.1. Theorem 4.1 of [**M2**] includes the hypothesis that there is a $L > 0$ such that β not L separated from α, a hypothesis not necessarily satisfied here. Accordingly, for any $L > 1$ let

$$S_L(\alpha) = \{\beta \in h \circ H^{-1}(\alpha) : \ \beta \text{ not } L \text{ separated from } \alpha \text{ on } Y_\beta\}.$$

Let $V_L(\alpha) = (h \circ H)^{-1}(\alpha) - S_L(\alpha)$. We consider first the set $V_L = \cup_\alpha V_L(\alpha)$. For any $\beta \in V_L$ let $\Gamma = (\gamma_1, \ldots, \gamma_p)$ be a system that L separates β from $\alpha = h \circ H(\beta)$ on Y_β. Assume $L > M = \sup |\alpha|_\beta$. Then $|\Gamma|_\beta \le \frac{|\alpha|_\beta}{L} \le M/L < 1$ so $\Gamma \ne \beta$ since $|\beta|_\beta = 1$. Further we can assume there is no Γ' separating β from Γ with $|\Gamma'|_\beta \le |\Gamma|_\beta/2$ for otherwise we could choose Γ' instead of Γ. Let $C_0 = 2^{C_1}(C_1)^{2C_1}$ where C_1 is the maximum number of saddle connections not intersecting at nonzeroes.

LEMMA 4.2. *There is a* $\gamma = \gamma_i \in \Gamma$ *such that* β *is not* C_0 *separated from* γ *on* Y_β.

PROOF: If β is C_0 separated from each $\gamma_i \in \Gamma$ by a Γ_i we could combine these Γ_i at most C_1 times using Lemma 3.3 to form Γ' 2 separating β from each $\gamma_i \in \Gamma$; that is; 2 separating β from Γ contradicting that there is no such Γ' on Y_β. $\square$

Now to each $\beta \in V_L$ associate to $H(\beta)$ the γ given by Lemma 4.2. That is, define

$$\gamma = g(H(\beta))$$

to be the map which associates to β of unit length, γ of length $\le M/L$ on Y_β.

Now suppose $\beta_1, \beta_2 \in V_L$ and $\gamma = g \circ H(\beta_1) = g \circ H(\beta_2)$. By Lemma 4.2 β_i is not C_0 separated from γ on Y_{β_i} but β_2 may be C_0 separated from γ on Y_{β_1}. However we have a lemma proved in [**M2**] which takes care of this situation.

LEMMA 4.3. *([**M2**, Lemma 3.2]).   For each* γ *there are sets* $B_1(\gamma), \ldots, B_p(\gamma)$, $p \le C_1$, *and a* $C_2 = C_2(C_0)$ *such that*

i) $\displaystyle\bigcup_{i=1}^{p} B_i(\gamma) = (g \circ H)^{-1}(\gamma)$;

ii) *for each* $i \le p$ *there is* $\beta_i \in B_i$ *so that* $\dfrac{|\beta|_{\beta_i}}{|\gamma|_{\beta_i}} \le \dfrac{8T}{\sigma_0|\gamma|}$ *for* T *sufficiently large for any* $\beta \in B_i$;

iii) *each* $\beta \in B_i$ *is not* C_2 *separated from* γ *on* Y_β.

Remark. The lemma says we can write $(g \circ H)^{-1}(\gamma)$ as a union of sets such that β is not C_2 separated on an appropriate Y_{β_i}; the β_i are chosen as in the discussion before Lemma 4.1.

LEMMA 4.4. *There is a constant $M_1 = M_1(M, \sigma_0)$ such that for any $\gamma \in g \circ H(V_L)$ we have $\epsilon \leq |\gamma| \leq \frac{M_1 T}{L}$. Recall ϵ is the shortest saddle connection on X.*

PROOF: After a Teichmüller map with dilatation $\sigma_0 T \leq K^{1/2} \leq T$ we have $|\gamma|_\beta \leq M/L$. But

$$\max\left(K^{1/2} h_{\gamma, \beta}, \frac{v_{\gamma, \beta}}{K^{1/2}} \right) \leq |\gamma|_\beta$$

so

$$v_{\gamma, \beta} \leq \frac{TM}{L} \quad \text{and} \quad h_{\gamma, \beta} \leq \frac{M}{L\sigma_0 T}.$$

$\square$

Proof of Theorem 2. By Theorem 4.1 of [**M2**] and Lemma 4.3 there are constants $M_2 = M_2(C_2)$, and p such that for any γ

$$\operatorname{card}(g \circ H)^{-1}(\gamma) \leq M_2 \frac{T}{|\gamma|} \left(\log \frac{T}{|\gamma|} \right)^p.$$

Choose k_1 so that

$$(4.1) \qquad \sum_{j \geq k_1} c_2 M_2 \sigma_0^j (\log \sigma_0^{-j})^p \leq c_3/3$$

where c_2, c_3 are the constants given in 2.1 and Corollary 2.2. This is possible since $\sum_{j=1}^{\infty} \sigma_0^j (j)^p$ converges. Now let $L = M_1 \sigma_0^{-k_1}$, M_1 given by Lemma 4.4. By Lemma 4.4 each γ satisfies $\epsilon \leq |\gamma| \leq \sigma_0^{k_1} T$. Divide the interval $[\epsilon, \sigma_0^{k_1} T]$ into intervals of the form $I_j = (\sigma_0^{j+1} T, \sigma_0^j T]$ $j \geq k_1$. By (2.1) the cardinality of the set of γ with lengths in I_j is $\leq c_2 \sigma_0^{2j} T^2$. Thus

$$\operatorname{card} V_L = \operatorname{card}\left(\bigcup_{\gamma \in (g \circ H)(V_L)} (g \circ H)^{-1}(\gamma) \right)$$

$$(4.2) \qquad \leq \sum_{j \geq k_1} M_2 c_2 \sigma_0^{2j} T^2 \, \sigma_0^{-j} (\log \sigma_0^{-j})^p$$

$$\leq c_3/3 \, T^2 \quad \text{by (4.1)}$$

Next we consider $S_L = \cup_\alpha S_L(\alpha)$. Exactly as in Lemma 4.4 we have $M_3 = M_3(M, \sigma_0)$ such that $|\alpha| \leq M_3 T$ for each $\alpha \in (h \circ H)(S_L)$. Recall for $\beta \in S_L$, by definition, β not L separated from $\alpha = h \circ H(\beta)$ on Y_β. Again applying Lemma 4.3 there is a $L' = L'(L)$ such that $(h \circ H)^{-1}(\alpha)$ can be written as a union of sets $B_i(\alpha)$ such that

i) for some $\beta_i \in B_i$, $\dfrac{|\beta|_{\beta_i}}{|\alpha|_{\beta_i}} \leq \dfrac{8T}{\sigma_0 |\alpha|}$ for all $\beta \in B_i$,

ii) each $\beta \in B_i$ is not L' separated from α on Y_{β_i}. Thus again by Theorem 4.1 of [**M2**] there is $M_4 = M_4(L')$ and p such that

$$\operatorname{card}(h \circ H)^{-1}(\alpha) \leq M_4 \frac{T}{|\alpha|} \left(\log \frac{T}{|\alpha|} \right)^p.$$

Now choose J so that

$$\sum_{j \geq J} c_2 M_4 \sigma_0^j (\log \sigma_0^{-j})^p \leq c_3/3$$

and divide $[\epsilon, M_3 T]$ into intervals $I_j = [\sigma_0^{j+1} T, \sigma_0^j T]$ as before. By (2.1)

$$\operatorname{card}(C(\sigma_0^j T) - C(\sigma_0^{j+1} T)) \leq \operatorname{card} C(\sigma_0^j T) \leq c_2 \sigma_0^{2j} T^2.$$

Thus

$$\operatorname{card}(h \circ H)^{-1}(C(\sigma_0^J T)) \leq \sum_{j \geq J} M_4 c_2 \sigma_0^{2j} T^2 \sigma_0^{-j} (\log \sigma_0^{-j})^p$$

(4.3)
$$\leq \frac{c_3 T^2}{3}$$

But we know card $S(T) \geq c_3 T^2$. Let

$$S'(T) = S(T) - V_L - (H \circ h)^{-1}(C(\sigma_0^J T)).$$

By (4.2) and (4.3) we have card $S'(T) \geq \frac{c_3 T^2}{3}$. But $S'(T)$ consists of those β such that

i) the associated $\alpha = h \circ H(\beta)$ has length at least $\sigma_0^J T$ on X.

ii) β is not L separated from α on Y_β.

After a Teichmüller map with maximal dilatation $K^{1/2} \leq T$, by i) above α has length at least $\frac{1}{2} v_0^J$ on Y_β.

By exactly the same argument as before,

$$\operatorname{card}(h \circ H)^{-1}(\alpha) \leq M_4 \frac{T}{|\alpha|} \left(\log \frac{T}{|\alpha|} \right)^p$$

where M_4 depends on L. Since $|\alpha| \geq \sigma_0^J T$,

$$\text{card}(h \circ H)^{-1}(\alpha) \leq M_5 \sigma_0^{-J} J^p$$

for some M_5. Since this cardinality is bounded independently of T for each such $\alpha \in h \circ H(S')$ and card $S' \geq c_3/3 \, T^2$ we must have

$$\text{card}((h \circ H)S') \geq c_4 T^2$$

for some $c_4 > 0$. But we have $\alpha \in C(MT)$ for each $\alpha \in h \circ H(S')$. Thus we have shown $C(MT) \geq c_4 T^2$ for fixed constant M, c_4. This concludes the proof.

Department of Mathematics, Statistics, and Computer Science, University of Illinois at Chicago, Chicago, Illinois 60680

REFERENCES

[**B**] Boshernitzan, M., *A condition for minimal interval exchange maps to be uniquely ergodic*, Duke Math. Journal **52** (1985), 723–752.

[**G**] Gutkin, E., *Billiards on almost integrable polyhedral surfaces*, Ergodic Th. and Dynam. Sys. **4** (1984), 569–584.

[**K**] Katok, A., *The growth rate for the number of singular and periodic orbits for a polygonal billiard*, preprint.

[**K-M-S**] Kerckhoff, S., Masur, H. and Smillie, J., *Ergodicity of billiard flows and quadratic differentials*, Annals of Math. **124** (1986), 293–311.

[**M1**] Masur, H., *Closed trajectories for quadratic differentials with an application to billiards*, Duke Math. Journal **53** (1986), 307–314.

[**M2**] __________, *The asymptotics of saddle connections for a quadratic differential*, to appear in Ergod. Th. and Dynamical Systems.

[**R**] Rees, M., *An alternative approach to the ergodic theory of measured foliations on surfaces*, Ergod. Th. and Dynamical Systems **1** (1981), 461–485.

[**St**] Strebel, K., "Quadratic differentials," Springer–Verlag, 1984.

Rational solutions of $\sum_{i=1}^{3} a_i x_i^2 = dx_1 x_2 x_3$ and simple closed geodesics on Fricke surfaces

BY MARK SHEINGORN

§1. Introduction.

Let H be the upper-half plane, $\Gamma(n) = \{\gamma \in SL(2, \mathbb{Z}) | \gamma \equiv \pm I \bmod n\}$, Γ' the commutator subgroup of $\Gamma(1)$ and $\Gamma^3 = \langle \gamma^3 | \gamma \in \Gamma(1) \rangle$. It has recently been established that the integral solutions of the Markov diophantine equation $x^2 + y^2 + z^2 = 3xyz$ characterize the simple closed geodesics on the modular surface $H/\Gamma(3)$, H/Γ', and H/Γ^3 ([H], [L-S], [S]). In this paper we seek analogous results for more general diophantine equations, specifically equations,

$$(1.0) \qquad \sum_{i=1}^{3} a_i x_i^2 = dx_1 x_2 x_3; \quad (a_1, a_2, a_3, d) = 1.$$

We find that *rational* solutions of any such equation lie in disjoint, essentially binary trees. (Actually the same trees exist when the a_i and d and the solutions are taken to be real. It is the number-theoretic bent of this paper, and the observation that the rational solutions of the (general) case (1.0) being the appropriate generalization of the integral solutions of the (special) case of Markov's equation that led us to stress the word *rational* in the above text.)

For a given equation, each tree is associated with a single Riemann surface of signature $[0; 2, 2, 2, \infty]$ and the solutions of (1.0) in that tree characterize the simple closed geodesics of that surface. From the solutions, there is an explicit algorithm to construct a Fuchsian group Φ_0 representing the surface and the hyperbolic matrices whose axes project to all simple closed geodesics on the surface. Also from our presentation of Φ_0, we obtain presentations of Ξ_0 and Ψ_0; these Fuchsian subgroups of Φ_0 represent surfaces of type $[1; \infty]$ and $[0; \infty, \infty, \infty, \infty]$ respectively. The same hyperbolic axes project to all simple closed geodesics on these surfaces also. (Surfaces with these signatures are called Fricke surfaces.) Finally, we give an explicit construction of all the rational trees of (1.0), concluding with some remarks about integral solutions.

This work was supported in part by NSF and PSC–CUNY Grants.

It turns out that there are six cases of (1.0), given by Rosenberger in [**R**] and described in the sequel, in which there is an entirely integral tree containing all integral solutions. We call these equations the Rosenberger cases. The Markov equation happens to be one of these and that fact may have disguised the role that the rationals play here.

The original observation of the relationship between Fricke's equation, the Markov equation and geodesics on punctured tori is due to H. Cohn [C, for example]. The proofs of the results in this paper depend heavily on the results given in Keen [**K**], Nielsen [**N**] and [**S**]. Given this machinery, the proofs herein are not difficult or deep. The paper's novelty, such as it is, lies in conceiving the nature and statement of the results.

Section 2 is a brief recapitulation of work on Fricke surfaces and Fricke's equation:

$$(1.1) \qquad\qquad x^2 + y^2 + z^2 = xyz.$$

We describe how positive solutions lie in disjoint trees, with each tree being associated with a certain Fricke surface. The way these solutions characterize simple closed geodesics on Fricke surfaces is also given in Section 2. (They are established in [**H**], [**K**], [**L-S**], and [**N**].) Section 3 considers integral solutions of (1.0), explaining the significance of the Rosenberger cases in this context and giving a few other examples that may indicate the lack of an elegant structure for only integral solutions.

A note on notations: we suppress the range of the index i, namely $i = 1, 2, 3$. So $X_i > 2$ applies to three variables and (r_i) is a 3-tuple.

§2. Nielsen Transformations, Fricke's Equation and Simple Closed Geodesics on Fricke Surfaces.

The *Nielsen transformations* of an ordered real triple are

$$(X, Y, Z) \;\to\; (X, XY - Z, Y)$$

$$(X, Y, Z) \;\to\; (Y, YZ - X, Z)$$

$$(X, Y, Z) \;\to\; (X, Z, XZ - Y)$$

If we start with a solution of (1.1), any Nielsen transformation gives us a solution of (1.1). It is easily checked that a Nielsen triple with $X_1 X_2 X_3 > 0$ has $X_i > 2$. Define the *height* of a positive solution (X_i) of (1.1) to be $\sum_{i=1}^{3} X_i$. The fact that $X_i > 2$ implies that at most one Nielsen transformation of a positive solution reduces its height. A positive solution

(X_0, Y_0, Z_0) of (1.1) is called *minimal* if no Nielsen transformation reduces its height. This amounts to having $X_0 \leq Y_0 Z_0/2$; $Y_0 \leq X_0 Z_0/2$, $Z_0 \leq X_0 Y_0/2$. Starting with any positive solution of (1.1), we may reduce its height by a sequence of Nielsen transformations (uniquely determined) until we reach a minimal solution (also unique). The possibility of an infinite regression may be ruled out thus: Suppose we had an infinite sequence of Nielsen triples (X_i, Y_i, Z_i) gotten by successive descending Nielsen transformations, $2 \leq X_i \leq Y_i \leq Z_i$. Then $X_i Y_i - Z_i \leq Y_i$ since otherwise $(X_{i+1}, Y_{i+1}, Z_{i+1})$ would be minimal and $2Z_i - X_i Y_i \to 0$ since the sequence of heights descends and hence tends to a limit. These two facts imply that $X_i \to 2$ and $Z_i - Y_i \to 0$, and this is a contradiction since the sequence is bounded and the Fricke equation has no solutions with $X = 2$, $Y = Z$. (The author is grateful to the referee for this argument.)

The Fuchsian group in question, a Fricke group, may be specified as follows. Let (A_0, B_0, C_0) be a minimal positive solution of (1.1). Assume $A_0 \leq B_0 \leq C_0 (\leq A_0 B_0/2)$. For $k > 0$, define $Q_0, R_0, S_0 (\in SL(2, \mathbb{R}))$ as

$$Q_0 = \begin{bmatrix} 0 & \frac{k}{A_0} \\ -\frac{A_0}{k} & 0 \end{bmatrix}, \quad R_0 = \begin{bmatrix} -B_0 + \frac{C_0}{A_0} & -k\left(B_0 - \frac{C_0}{A_0} - \frac{B_0}{A_0^2}\right) \\ \frac{B_0}{k} & B_0 - \frac{C_0}{A_0} \end{bmatrix},$$

$$S_0 = \begin{bmatrix} \frac{B_0}{A_0} & k\left(\frac{B_0}{A_0} - \frac{C_0}{A_0^2}\right) \\ -\frac{C_0}{k} & -\frac{B_0}{A_0} \end{bmatrix}.$$

Then the associated Fricke group $\Phi_0 = \Phi_0(A_0, B_0, C_0, k) = \langle Q_0, R_0, S_0 \rangle$. The group has signature $(0; 2, 2, 2, \infty)$ and any Fuchsian group with that signature is conformally equivalent to exactly one of these. (The matrix $Q_0 S_0 R_0 = \begin{pmatrix} 1 & k \\ 0 & 1 \end{pmatrix}$ specifies the puncture.

Corresponding to the tree of solutions of (1.1) stemming from a minimal solution (a binary tree, except for the first branch) there is a tree of *hyperbolic* triples which are generators of the group Φ_0, beginning with $(T_0, U_0, V_0) =: (S_0 R_0, -Q_0 S_0, U_0 T_0^{-1})$ [Sc; **K**, Theorem 4 for $(1; \infty)$ case]. These hyperbolics are exactly the ones with axes projecting to simple closed geodesics on H/Φ_0 [**N** for $(1; \infty)$ case; **S**]. It can be shown [**S**] that these axes project to precisely the simple closed geodesics on H/Ξ_0 and H/Ψ_0, surfaces of signatures $(1; \infty)$ and $(0; \infty, \infty, \infty, \infty)$, where Ξ_0 and Ψ_0 are explicitly defined subgroups of Φ_0 and $[\Phi_0 : \Xi_0] = 2$; $[\Phi_0 : \Psi_0] = 4$. Moreover, every surface with one of the latter signatures is conformally equivalent to one of these "subgroup of Φ_0" surfaces, for exactly one choice of the parameters A_0, B_0, C_0, k.

To briefly recapitulate, trees of solutions of Fricke's equation (1.1) may be used to generate the set of simple closed geodesics on any surface of signature $[0; 2, 2, 2, \infty]$ or $[1; \infty]$ or $[0; \infty, \infty, \infty, \infty]$.

§3. The Geometry of Rational Solutions of $\sum_{i=1}^{3} a_i x_i^2 = d x_1 x_2 x_3$.

We begin with a rational solution (r_i) of the equation (1.0). This is equivalent to starting with

$$(3.1) \qquad \delta(r_i \sqrt{a_i}), \quad \delta = \frac{d}{\sqrt{a_1 a_2 a_3}}$$

as a solution of Fricke's equation (1.1). By using Nielsen transformations, we can recover the minimal element generating the tree of (3.1). It is easy to check that everything in this tree has the form (3.1), that is, each triple's entry may be written $\delta \cdot \text{rational} \cdot \sqrt{a_i}$. In this way, a rational solution of (1.0) may be used to obtain a tree of rational solutions of (1.0). (Note that an *ordered* minimal triple (r_1, r_2, r_3) with $r_1 \leq r_2 \leq r_3$ generates solutions of (1.0) only if the replaced member of the triple under the Nielsen transformation occupies the same position. For example $(r_1, r_2, \frac{d}{a_3} r_1 r_2 - r_3)$ is a solution but $(r_1, \frac{d}{a_3} r_1 r_2 - r_3, r_2)$ may not be. The Markov equation is symmetric in x, y, and z and so this consideration doesn't arise in that case.) Because these trees are uniquely determined by their (unique) minimal element, these trees are disjoint. Moreover, each tree specifies one Fricke surface of signature $[0; 2, 2, 2, \infty]$ [and one each of type $[1; \infty]$ and $[0; \infty, \infty, \infty, \infty]$]] with the nodes (solutions) of the tree giving the simple closed geodesics on the surface(s).

It is not difficult to construct all rational solutions of (1.0). Begin with any triple of (say) integers (ρ_i), then

$$(3.2) \qquad (\rho_i \tau), \quad \tau = \frac{\sum_{i=1}^{3} a_i \rho_i^2}{d \rho_1 \rho_2 \rho_3}$$

is a rational solution of (1.0). It is also easy to see that (3.2) is minimal if and only if

$$(3.3) \qquad a_i \rho_i^2 + a_j \rho_j^2 - a_k \rho_k^2 \geq 0,$$

for all permutations (i, j, k) of $(1, 2, 3)$. (Here we assume $\rho_1 \rho_2 \rho_3 > 0$ and there is a slight abuse of language: (3.2) is minimal means that the corresponding solution of Fricke's equation given in (3.1) is minimal.)

EXAMPLE 1: Consider $x^2 + y^2 + z^2 = 3xyz$. Here $\rho_1 = 1$, $\rho_2 = 2$, $\rho_3 = 5$ gives $\tau = 1$. This is not minimal as $(1,1,1)$ is the minimal source of this tree, the famous Markov tree. Now $\rho_1 = 2$, $\rho_2 = 2$, $\rho_3 = 1$ gives $\tau = 3/4$ and $(\rho_i \tau) = (3/2, 1/2, 1/4)$ is also a solution of $x^2 + y^2 + z^2 = 3xyz$. Verifying (3.3) shows that it is minimal and thus the source of a tree of relational solutions, disjoint from the famous Markov tree, whose nodes give the simple closed geodesics on a Riemann surface of type $[0; 2, 2, 2, \infty]$ which is conformally different from the Markov case H/Γ^3.

Obviously, if we require that $(\rho_1, \rho_2, \rho_3) = 1$ then the representation (3.2) of a rational solution (r_i) exists and is unique. Thus beginning with equation (1.0) and an enumeration of the relatively prime integral triples we can recursively enumerate all minimal rational solutions $(\rho_i \tau)$ and with them *all* rational solutions, segregated by trees whose Riemann surfaces have their simple closed geodesics described by the rational solutions.

§4. **Integral Solutions of $\sum_{i=1}^{3} a_i x_i^2 = d x_1 x_2 x_3$.**

First, note that the solution preserving Nielsen transformations given in §1 amount to:

$$(r_i) \; \to \; \left(r_1, r_2, \frac{d r_1 r_2}{a_3} - r_3 \right)$$

(4.1)
$$(r_i) \; \to \; \left(r_1, \frac{d r_1 r_3}{a_2} - r_2, r_3 \right)$$

$$(r_i) \; \to \; \left(\frac{d r_2 r_3}{a_1} - r_1, r_2, r_3 \right).$$

If $a_i \mid d$, $\forall a_i$ (4.1) shows that if (r_i) is an *integral* solution, then all solutions in its tree are integral. If, additionally, we can show that (1.0) has a *unique minimal integral* solution, then all integral solutions must lie in this one tree, because of disjointness.

Rosenberger, in [**R**], determines all cases *with $a_i \mid d$* having a unique minimal integral solution. These cases, which we call the Rosenberger cases, are:

$$x^2 + y^2 + x^2 = xyz$$

$$x^2 + y^2 + z^2 = 3xyz$$

$$x^2 + y^2 + 2z^2 = 2xyz$$

$$x^2 + y^2 + 2z^2 = 4xyz$$

$$x^2 + y^2 + 5z^2 = 5xyz$$

$$x^2 + 2y^2 + 3z^2 = 5xyz.$$

In general, the situation is not so simple. Rational trees may have some integral solutions on them.

EXAMPLE 2: $x^2 + y^2 + kz^2 = (k+2)xyz$ has the solution $(1,1,1)$. This is minimal only if $k = 1$ or 2, and these are Rosenberger cases. If $k > 2$, one Nielsen transformation brings us to $(1,1,2/k)$ which is minimal but not integral.

In this example, because a_1 and a_2 divide d, we could do Nielsen transformations discarding the first and second elements successively and obtain an entirely *integral path* of solutions:

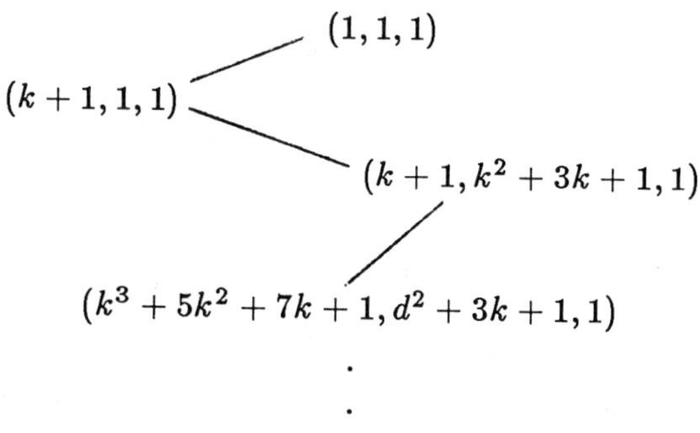

An integral path can be obtained any time a_i, $a_j | d$ and we possess (r_i), an *integral* solution.

A further complication is that for certain integral trees two apparently unrelated equations may coincide.

EXAMPLE 3: $2x^2 + 3y^2 + 4z^2 = 6xyz$. $(2,2,1)$ is a solution of this equation and it is minimal. It is easy to show that every solution stemming from $(2,2,1)$ has the form $(2\ell, 2\ell_2, \ell_3)$ for *integral* ℓ_1, ℓ_2, ℓ_3 substituting in the original equation yields: $\ell_3^2 + 2\ell_1^2 + 3\ell_2^2 = 6\ell_3\ell_1\ell_2$, a Rosenberger case.

I cannot resolve the question of whether more than one tree may contain
an integral solution. I see nothing to prevent this, but I have been unable
to find an example.

School of Mathematics, Institute for Advanced Study, Princeton, NJ 08540
Current Address: Dept. of Mathematics, Baruch College–CUNY, New York, NY 10010

REFERENCES

[C] Cohn, H., *Minimal Geodesics on Fricke's Torus-Covering*, Ann. of Math Studies **97** (1981), 73–86.

[H] Haas, A., *Diophantine approximation on hyperbolic Riemann surfaces*, Acta Math. (Uppsala) **156** (1986), 33–82.

[K] Keen, L., *On Fricke Moduli*, Ann. of Math. Studies **66** (1971), 205–224.

[L-S] Lehner, J. and Sheingorn, M., *Simple closed geodesics on $H^+/\Gamma(3)$ arise from the Markov spectrum*, BAMS (2) **11** (1984), 359–362.

[N] Nielsen, J., *Unter. zur Topoligie der geschlossenen zweiseitigen Flächen*, Acta Math. (Uppsala) **50** (1927), 189–358.

[R] Rosenberger, G., *Uber die Diophantische Gleichung $ax^2 + by^2 + cz^2 = dxyz$*, J. Riene Angew. für Math. **305** (1979), 122–125.

[Sc] Schmidt, A.L., *Minimum of quadratic forms with respect to Fuchsian groups I*, J. Riene Angew. für Math. **286/287** (1976), 341–368.

[S] Sheingorn, M., *Characterization of simple closed geodesics on Fricke surfaces*, Duke Math. J. **52** (1985), 535–545.

The period matrices of compact continuations of an open Riemann surface of finite genus

Dedicated to Professor Tadashi Kuroda on his sixtieth birthday

BY M. SHIBA

Introduction.

The purpose of the present paper is to study the space of compact continuations of an open Riemann surface of finite genus in connection with the Torelli space or the Siegel upper half space.

We will not give here the precise definition (see section 1.2) of compact continuations, but illustrate our results by considering special cases. To do so, let there be given an open Riemann surface R of genus $g(\geq 1)$, together with a canonical homology basis χ modulo dividing cycles. If a closed Riemann surface R' of genus g contains R as an open subset, χ gives rise to a canonical homology basis χ' of R'. Let (R', χ') generically denote such a pair. It then defines a "Torelli surface" (cf. [14], [25], for instance), or a point in the Torelli space of genus g. Denote by $C^*(R, \chi)$ the set of all such Torelli surfaces (R', χ'). Since, by Torelli's theorem ([5]), the Siegel upper half space $\mathfrak{S}_g$ of genus g well reflects the structure of the Torelli space, we may consider $\mathfrak{S}_g$ to study the space $C^*(R, \chi)$. Namely, we consider the (normalized) period matrix $\tau^*(R', \chi')$ of $(R', \chi') \in C^*(R, \chi)$.

To make the exposition concise, we shall restrict ourselves to the study of the diagonal $\Delta^*(R', \chi')$ of $\tau^*(R', \chi')$. The non-diagonal entries will be discussed elsewhere. One of our results reads: There exists a closed polydisk P in $\mathbb{C}^g$ such that (i) $P \supset \Delta^*(R', \chi')$ for any $(R', \chi') \in C^*(R, \chi)$, and (ii) $\Delta^*(R', \chi') \in \partial P$ if and only if (R', χ') is a canonical hydrodynamic

1980 Mathematics Subject Classification (1985 Revision). Primary 30Fxx, 30C70, 32G20; Secondary 14H15, 30C35, 32G15.

Key words and phrases. Continuation (extension, prolongation) of Riemann surfaces, hydrodynamic continuation, distinguished differential, complex velocity potential, period matrix, Torelli space, Siegel upper half space, extremal slit mapping, span, O_{AD}.

Part of this work was done during the author's stay at University of Hannover, FRG. The author is very grateful to Professors H. Tietz, E. Mues, G. Schmieder (now at University of Würzburg) and other colleagues for their warm hospitality and stimulative seminars. He also thanks DAAD (Deutscher Akademischer Austauschdienst) for the financial support.

continuation of (R, χ). Here, (R', χ') is called a canonical hydrodynamic continuation of (R, χ) if we can find, in the canonical basis for the holomorphic differentials on (R', χ'), some $d\Phi$ such that $\Phi | R$ is an S-function on R. The above result also shows that it suffices, for the discussion of the uniqueness of (R', χ'), to consider only the diagonal entries of the period matrices.

Our idea for the proof is basically classical, although the non-classical notion of hydrodynamic continuations also plays an important role throughout the paper. Extremal properties shared by hydrodynamic continuations can be compared to the well-known classical ones shared by, e.g., the minimal slit plane regions. Cf. [3], [9], [16] and [17].

It would be useful to mention the simplest case. If $g = 1$, the set $\{\tau^*(R', \chi') | (R', \chi') \in C^*(R, \chi)\}$ is precisely a closed disk in the upper half plane (see [4] and [20]). Since the Teichmüller space and the Torelli space coincide with each other for $g = 1$, the results in this paper can be recognized as a generalization of our preceding work [20].

The author would like to express his hearty thanks to the organizing committee of the workshop for their kindness and generosity of including my talk in the workshop and the present paper in the proceedings.

1. Compact continuations.

1.1. Let R be an open Riemann surface of finite genus $g(\geq 1)$ and $\chi = \{a_j, b_j\}_{j=1}^g$ a canonical homology basis of R modulo dividing cycles ([1]). The pair (R, χ) is fixed throughout this paper.

By a *realization* of (R, χ) we shall understand a triple (R', χ', I'), where R' is a closed Riemann surface of genus g, $\chi' = \{a_j', b_j'\}_{j=1}^g$ a canonical homology basis of R', and I' a conformal injection of R into R' such that $I'(a_j)$ and $I'(b_j)$ are respectively homologous to a_j' and b_j' $(j = 1, 2, \ldots, g)$. Two realizations (R', χ', I') and (R'', χ'', I'') of (R, χ) will be said to be *equivalent*, if there exists a conformal mapping f of R' onto R'' with $f \circ I' = I''$ (on R). Each equivalence class will be called a *compact continuation* of (R, χ). The compact continuation determined by a realization (R', χ', I') will be denoted by $[R', \chi', I']$, and the space of compact continuations of (R, χ) by $C(R, \chi)$.

1.2. If two realizations (R', χ', I') and (R'', χ'', I'') of (R, χ) are equivalent, then (R', χ') and (R'', χ'') obviously define the same "Torelli surface", or the same point in the Torelli space $\mathfrak{T}_g$ of genus g (cf. [14], [25]). If this is the case, (R', χ') and (R'', χ'') have the same "normalized period matrix" (see [22], p. 114 or [14]), so that we denote it by $\tau[R', \chi', I']$. As is well known, $\tau[R', \chi', I']$ belongs to $\mathfrak{S}_g$, the Siegel upper half space of genus g (see [2], p. 280 and [22], p. 129). We note that $\tau : C(R, \chi) \to \mathfrak{S}_g$ is not always injective as was shown in [20] for $g = 1$.

For later use, we shall give here the explicit form of $\tau[R', \chi', I']$. To this end set $\chi' = \{a'_j, b'_j\}_{j=1}^g$ and let $\{\varphi'_j\}_{j=1}^g$ be the canonical basis for the holomorphic differentials (cf. e.g., [23], p. 255) with respect to χ'. Namely, suppose $\int_{a'_k} \varphi'_j = \delta_{jk}$ for $j, k = 1, 2, \ldots, g$, where δ_{jk} is the Kronecker symbol. Then $\tau[R', \chi', I']$ is given by $(\tau'_{jk})_{j,k=1,2,\ldots,g}$ with $\tau'_{jk} = \int_{b'_k} \varphi'_j$ for $j, k = 1, 2, \ldots, g$. To make the dependence of τ'_{jk} on $[R', \chi', I']$ clear, we write $\tau'_{jk} = \tau_{jk}[R', \chi', I']$, $j, k = 1, 2, \ldots, g$.

1.3. Let $A = (\alpha_1, \alpha_2, \ldots, \alpha_g) \in \mathbb{R}^g$, and $[R', \chi', I'] \in C(R, \chi)$. Then there exists a unique holomorphic differential φ'_A on R' with $\int_{a'_j} \varphi'_A = \alpha_j$, $j = 1, 2, \ldots, g$. We set

$$\beta'_j(A) = \int_{b'_j} \varphi'_A, \quad j = 1, 2, \ldots, g,$$

and

$$\tau_A[R', \chi', I'] = (\beta'_1(A), \beta'_2(A), \ldots, \beta'_g(A)).$$

If we set $A_k = (\delta_{k1}, \delta_{k2}, \ldots, \delta_{kg})$, $k = 1, 2, \ldots, g$, we can easily see

$$\varphi'_{A_k} = \varphi'_k \quad (k = 1, 2, \ldots, g); \quad \varphi'_A = \sum_{j=1}^g \alpha_j \varphi'_j;$$

$$\beta'_j(A_k) = \tau'_{kj} \quad (j, k = 1, 2, \ldots, g),$$

$$\tau_{A_k}[R', \chi', I'] = (\tau'_{k1}, \tau'_{k2}, \ldots, \tau'_{kg}) \quad (k = 1, 2, \ldots, g).$$

We usually abbreviate $\tau_{A_k}[R', \chi', I']$ to $\tau_k[R', \chi', I']$.

2. Hydrodynamic continuations.

2.1. Let $t \in (-1, 1] := \{x \in \mathbb{R} \mid -1 < x \leq 1\}$ and F a (single- or multiple-valued) meromorphic function on R. We call F an S_t-*function* ([19], [20]), if $\mathrm{Im}\left(e^{-\frac{1}{2}\pi i t} dF\right)$ is distinguished in the sense of Ahlfors. When we do not need to refer explicitly to t, we call F simply an S-function. For the definition of distinguished differentials, see [1], p. 313. The name S-function originates from the German word "Strömungsfunktion" (see [7], p. 484). It should be noted, however, that "Strömungsfunktion" does not mean the stream function in fluid dynamics but means the complex velocity potential function (of a stationary flow of an ideal fluid on R).

2.2. The following theorem is fundamental to the present study. For the proof of this theorem, see [**19**], [**21**]. Cf. also [**20**].

THEOREM 1. *For any non-constant S_t-function F on R there exists a realization (R^F, χ^F, I^F) with the following properties:*

(I) *The (Lebesgue) area of $R^F \backslash I^F(R)$ vanishes.*

(II) *The transplant of dF via $(I^F)^{-1}$ extends to a meromorphic differential $d\tilde{F}$ on R^F.*

(III) *$\tilde{F}$ has a single-valued holomorphic branch on (a neighborhood of) $R^F \backslash I^F(R)$, and $\operatorname{Im}\left(e^{-\frac{1}{2}\pi it}\tilde{F}\right)$ takes on a constant value on each componenet of $R^F \backslash I^F(R)$.*

After its physical meaning (cf. [**20**], [**21**]), we call (R^F, χ^F, I^F) a *hydrodynamic realization* and $[R^F, \chi^F, I^F]$ a *hydrodynamic continuation* of (R, χ) with respect to the S-function F. Neither (R^F, χ^F, I^F) nor $[R^F, \chi^F, I^F]$ is determined uniquely by a given R, χ, and F in general (see [**20**]). Hence we consider, instead of individual $[R^F, \chi^F, I^F]$, the subspace $C(R, \chi; F)$ consisting of all hydrodynamic continuations of (R, χ) with respect to F.

3. Lemmas.

3.1. The Dirichlet norm of a holomorphic differential φ on R is denoted by $\|\varphi\|_R$. If we write $\varphi = d\Phi = \Phi'(z)dz$ in terms of a generic local parameter $z = x + iy$ on R, we have

$$\|\varphi\|_R = \left(2 \iint_R |\Phi'(z)|^2 dxdy\right)^{\frac{1}{2}}.$$

The *real* Dirichlet inner product of two holomorphic differentials $\varphi = d\Phi$ and $\psi = d\Psi$ on R with a finite Dirichlet norm is defined by

$$\langle \varphi, \psi \rangle_R = 2 \operatorname{Re} \iint_R \Phi'(z)\overline{\Psi'(z)}dxdy.$$

We recall that the differential of a holomorphic S-function has a finite Dirichlet norm. The following lemma can be proved without difficulty (cf. [**20**], Lemma 5).

LEMMA 1. *(a) Let Φ^o be a holomorphic S_o-function on R and set $\varphi^o = d\Phi^o$. Then*

$$\|\varphi^o\|_R^2 = -2 \sum_{j=1}^{g} \operatorname{Im}\left(\int_{a_j} \varphi^o \int_{b_j} \overline{\varphi^o}\right).$$

(b) Let φ^o be as in (a). If $\varphi = d\Phi$ is another holomorphic semiexact differential on R with $\|\varphi\|_R < \infty$, then

$$\langle \varphi, \varphi^o \rangle_R = 2 \sum_{j=1}^{g} \left\{ \int_{a_j} \operatorname{Re} \varphi \int_{b_j} \operatorname{Im} \varphi^o - \int_{b_j} \operatorname{Re} \varphi \int_{a_j} \operatorname{Im} \varphi^o \right\}.$$

3.2. Using the classical argument (cf. e.g., [**2**], pp. 56–61) *mutatis mutandis*, we have

LEMMA 2. *For each $k = 1, 2, \ldots, g$ and $t \in (-1, 1]$ there exists a holomorphic S_t-function Φ_k^t on R with $\int_{a_j} d\Phi_k^t = \delta_{jk}$, $j = 1, 2, \ldots, g$. The differential $d\Phi_k^t$ is uniquely determined.*

The system $\{\Phi_k^t\}_{k=1}^{g}$ may be called a *canonical basis* for the holomorphic S_t-functions on R. Cf. section 1.2. The following lemma is an immediate corollary of Lemma 2.

LEMMA 2'. *For any $A = (\alpha_1, \alpha_2, \ldots, \alpha_g) \in \mathbb{R}^g$ there exists an S_t-function Φ_A^t on R with $\int_{a_j} d\Phi_A^t = \alpha_j$, $j = 1, 2, \ldots, g$.*

As in section 1.3, we have $d\Phi_{A_k}^t = d\Phi_k^t$ $(k = 1, 2, \ldots, g)$ and $d\Phi_A^t = \sum_{k=1}^{g} \alpha_k d\Phi_k^t$. If we set $\tau_{kj}^t = \int_{b_j} d\Phi_k^t$ $(j, k = 1, 2, \ldots, g)$, the b_j-period $\beta_j^t(A)$ of $d\Phi_A^t$ is given by $\sum_{k=1}^{g} \alpha_k \tau_{kj}^t$, $j = 1, 2, \ldots, g$.

4. Extremal properties of hydrodynamic continuations.

4.1. Let $A = (\alpha_1, \alpha_2, \ldots, \alpha_g) \in \mathbb{R}^g$ and $t \in (-1, 1]$ be fixed in this section. Furthermore, let $[R', \chi', I']$ be any point in $C(R, \chi)$ and set $\chi' = \{a_j', b_j'\}_{j=1}^{g}$. Using the b_j'-periods $\beta_j'(A)$ of the holomorphic differential $\varphi' = \varphi_A'$ on R' (see section 1.3), we define $\pi_A^t : C(R, \chi) \to \mathbb{R}$ by

$$\pi_A^t[R', \chi', I'] = \operatorname{Im}\left(e^{-\pi i t} \sum_{j=1}^{g} \alpha_j \beta_j'(A) \right).$$

Obviously π_A^t is conatant on $C(R, \chi; \Phi_A^t)$.

4.2. We shall now show that $C(R, \chi; \Phi_A^t)$ minimizes π_A^t. More precisely, we have

THEOREM 2. *Let $A \in \mathbb{R}^g$ and $t \in (-1,1]$ be fixed as above, and let $[\tilde{R}, \tilde{\chi}, \tilde{I}] \in C(R, \chi; \Phi_A^t)$. Then*

$$\pi_A^t[R', \chi', I'] \geq \pi_A^t[\tilde{R}, \tilde{\chi}, \tilde{I}]$$

for all $[R', \chi', I'] \in C(R, \chi)$. Equality holds if and only if $[R', \chi', I'] \in C(R, \chi; \Phi_A^t)$.

PROOF: Let $[R', \chi', I']$, $\chi' = \{a_j', b_j'\}_{j=1}^g$ and φ' be as in the preceding section and let $d\Phi'$ be the transplant of φ' via I'. Then, by a well-known fact in the classical theory (cf. Lemma 1 (a); cf. [2] and [23], too), we have

$$\|d\Phi'\|_R^2 \leq \|\varphi'\|_{R'}^2 = -2 \sum_{j=1}^g \text{Im} \left(\int_{a_j'} \varphi' \int_{b_j'} \overline{\varphi'} \right) = 2 \sum_{j=1}^g \alpha_j \, \text{Im} \, \beta_j'(A).$$

Next, take an arbitrary $[\tilde{R}, \tilde{\chi}, \tilde{I}] \in C(R, \chi; \Phi_A^t)$ and set $\tilde{\chi} = \{\tilde{a}_j, \tilde{b}_j\}_{j=1}^g$. If $\tilde{\varphi} = \tilde{\varphi}_A$ is the holomorphic differential on the closed surface $\tilde{R}$ with $\int_{\tilde{a}_j} \tilde{\varphi} = \alpha_j$, $j = 1, 2, \ldots, g$, then the transplant of $\tilde{\varphi}$ via $\tilde{I}^{-1}$ coincides with $d\Phi_A^t$ by Theorem 1, Lemma 2' and Lemma 1 (a). Consequently, we have $\int_{\tilde{b}_j} \tilde{\varphi} = \beta_j^t(A)$, $j = 1, 2, \ldots, g$, so that we have by Lemma 1 (a)

$$\|d\Phi_A^t\|_R^2 = -2 \sum_{j=1}^g \text{Im} \left(\int_{a_j} d\Phi_A^t \int_{b_j} \overline{d\Phi_A^t} \right) = 2 \sum_{j=1}^g \alpha_j \, \text{Im} \, \beta_j^t(A).$$

Since $e^{-\frac{1}{2}\pi i t} \Phi_A^t$ is an S_o-function on R, we have by Lemma 1 (b)

$$\langle d\Phi', d\Phi_A^t \rangle_R$$
$$= 2 \sum_{j=1}^g \alpha_j \left[\cos \frac{1}{2}\pi t \, \text{Im} \left(e^{-\frac{1}{2}\pi i t} \beta_j^t(A) \right) + \sin \frac{1}{2}\pi t \, \text{Re} \left(e^{-\frac{1}{2}\pi i t} \beta_j'(A) \right) \right]$$
$$= \sum_{j=1}^g \alpha_j \left[\sin \pi t \, \text{Re} \left(\beta_j'(A) - \beta_j^t(A) \right) - \right.$$
$$\left. - \cos \pi t \, \text{Im} \left(\beta_j'(A) - \beta_j^t(A) \right) + \text{Im} \left(\beta_j'(A) + \beta_j^t(A) \right) \right]$$
$$= \sum_{j=1}^g \alpha_j \, \text{Im} \left[e^{-\pi i t} \{\beta_j^t(A) - \beta_j'(A)\} \right] + \sum_{j=1}^g \alpha_j \, \text{Im} \left[\beta_j'(A) + \beta_j^t(A) \right].$$

Consequently, we have

$$0 \leq \|d\Phi' - d\Phi_A^t\|_R^2 = \|d\Phi'\|_R^2 + \|d\Phi_A^t\|_R^2 - 2\langle d\Phi', d\Phi_A^t \rangle_R$$
$$\leq 2 \sum_{j=1}^g \alpha_j \, \text{Im} \left[e^{-\pi i t} \{\beta_j'(A) - \beta_j^t(A)\} \right]$$
$$= 2\{\pi_A^t[R', \chi', I'] - \pi_A^t[\tilde{R}, \tilde{\chi}, \tilde{I}]\},$$

which is to be proved.

5. Canonical hydrodynamic continuations.

5.1. We shall call a point $[R', \chi', I'] \in C(R, \chi)$ a *canonical hydrodynamic continuation* of (R, χ) if it belongs to $C(R, \chi; \Phi_k^t)$ for some integer k $(1 \leq k \leq g)$ and $t \in (-1, 1]$. In section 5.2 we shall characterize the space of canonical hydrodynamic continuations of (R, χ). For the time being, we keep an integer k fixed and let t vary in the interval $(-1, 1]$. Then we have, for any $[\tilde{R}, \tilde{\chi}, \tilde{I}] \in C(R, \chi; \Phi_k^t)$,

$$\tau_k[\tilde{R}, \tilde{\chi}, \tilde{I}] = (\tau_{k1}^t, \tau_{k2}^t, \ldots, \tau_{kg}^t).$$

It should be noted, however, that the corresponding equality for $\ell \neq k$ does not hold in general.

By Lemma 2 the differential $e^{-\frac{1}{2}\pi it} d\Phi_k^t$ is represented as a linear combination of $d\Phi_k^o$ and $id\Phi_k^1$ with real coefficients. In fact, we can easily see

$$e^{-\frac{1}{2}\pi it} d\Phi_k^t = \cos \frac{1}{2}\pi t \; d\Phi_k^o - i \sin \frac{1}{2}\pi t \; d\Phi_k^1.$$

Computing the b_j-periods of both sides, we obtain

$$\tau_{kj}^t = \frac{1}{2}(\tau_{kj}^o + \tau_{kj}^1) + \frac{1}{2}(\tau_{kj}^o - \tau_{kj}^1)e^{\pi it}, \quad j = 1, 2, \ldots, g.$$

Hence we have proved

THEOREM 3. *Every τ_{kj}^t moves on a circle when t varies in the interval $(-1, 1]$. More precisely:*

(1)
$$\tau_{kj}^t = \tau_{kj}^* + \rho_{kj}e^{(t - t_{kj})\pi i}, \quad t \in (-1, 1],$$

where

$$\tau_{kj}^* = \frac{1}{2}(\tau_{kj}^1 + \tau_{kj}^o), \quad \rho_{kj} = \frac{1}{2}|\tau_{kj}^1 - \tau_{kj}^o|,$$

and

$$t_{kj} = 1 - \frac{1}{\pi} \arg(\tau_{kj}^1 - \tau_{kj}^o) \in (-1, 1].$$

5.2. We shall now study exclusively the behavior of the diagonal elements of the (normalized) period matrix $\tau[R', \chi', I']$ when $[R', \chi', I']$ runs through $C(R, \chi)$. To this end, fix an integer k $(1 \leq k \leq g)$ and $t \in (-1, 1]$ as before. The following proposition is then an easy consequence of Theorem 2.

PROPOSITION 1. *For any* $[R', \chi', I'] \in C(R, \chi)$ *and* $t \in (-1, 1]$

(2) $$\mathrm{Im}\left[e^{-\pi i t}\{\tau_{kk}[R', \chi', I'] - \tau_{kk}^t\}\right] \geq 0;$$

the equality holds if and only if $[R', \chi', I'] \in C(R, \chi; \Phi_k^t)$. *In particular,*

(3) $$\mathrm{Im}\,\tau_{kk}^o \leq \mathrm{Im}\,\tau_{kk}[R', \chi', I'] \leq \mathrm{Im}\,\tau_{kk}^1.$$

Inequality (3), together with Theorem 3, now implies

PROPOSITION 2. $t_{kk} = \frac{1}{2}$ *for* $k = 1, 2, \ldots, g$.

Substituting equation (1) into inequality (2), we have

$$\mathrm{Im}\left[e^{-\pi i t}\{\tau_{kk}[R', \chi', I'] - \tau_{kk}^*\}\right] \leq \rho_{kk}.$$

Since the parameter t can be arbitrarily chosen in the interval $(-1, 1]$, it follows immediately from the above inequality that

$$|\tau_{kk}[R', \chi', I'] - \tau_{kk}^*| \leq \rho_{kk}.$$

We have hence proved the following theorem, which characterizes the space of canonical hydrodynamic continuations.

THEOREM 4. *Let* $\Delta : C(R, \chi) \to \mathbb{C}^g$ *be defined by*

$$\Delta[R', \chi', I'] = (\tau_{11}', \tau_{22}', \ldots, \tau_{gg}')$$

with $\tau_{kk}' = \tau_{kk}[R', \chi', I'], k = 1, 2, \ldots, g$ *(see section 1.3). Then,* $\Delta(C(R, \chi))$ *is contained in the closed polydisk*

$$P := \prod_{k=1}^{g} \{|\tau_{kk} - \tau_{kk}^*| \leq \rho_{kk}\}.$$

If $[R', \chi', I']$ *is a canonical hydrodynamic continuation, then* $\Delta[R', \chi', I']$ *lies on the boundary of* P, *and vice versa.*

5.3. If $\rho_{kk} = 0$ for some k $(1 \leq k \leq g)$, then by Theorem 4 $C(R, \chi)$ consists of a single point. Hence, by a theorem of Mori (**[11]**), we know that R belongs to the class O_{AD}. On the other hand, by a theorem due to Oikawa, (**[13]**), $R \in O_{AD}$ implies that $\rho_{jj} = 0$ for all $j = 1, 2, \ldots, g$. Cf. also **[15]**. Thus we have proved

THEOREM 5. *The following four conditions are equivalent to one another:*

(a) $\rho_{kk} = 0$ *for some* k $(1 \le k \le g)$.

(b) $\rho_{jj} = 0$ *for every* $j = 1, 2, \ldots, g$.

(c) R *is of class* O_{AD}.

(d) $C(R, \chi)$ *consists of a single point.*

In particular, any ρ_{kk} can be used as a kind of generalized span. See [20]. Cf. also [16], [17], and [18].

5.4. As we announced earlier, we have not fully discussed the non-diagonal entries of the period matrices of points in $C(R, \chi)$. As for the hydrodynamic continuations, however, Theorem 3 gives rather complete description of the behavior of non-diagonal entries. In a forthcoming paper we shall prove that every entry $\tau_{jk}[R', \chi', I']$ is a bounded function of $[R', \chi', I']$ on $C(R, \chi)$. This result is closely related with theorems of Heins ([6]) and of Oikawa ([12]). Cf. [8] and [24], too.

ADDED IN PROOF: By a theorem of Oikawa ([13]) or by our reasoning above, we see that condition (d) in Theorem 5 can be replaced by any one of the following two conditions:

(d_1) R admits a unique continuation in the sense of Nevanlinna (see [11]).

(d_2) R admits a unique continuation in the sense of Oikawa ([13]).

Department of Mathematics, Hiroshima University, Hiroshima, Japan

References

1. Ahlfors, L.V. and Sario, L., "Riemann surfaces," Princeton Univ. Press, Princeton, 1960, 382pp.
2. Farkas, H.M. and Kra, I., "Riemann surfaces," Springer, New York–Heidelberg–Berlin, 1980, 337pp.
3. Grötzsch, H., *Über das Parallelschlitztheorem der konformen Abbildung schlichter Bereiche*, Ber. Verh. Sächs. Akad. Wiss. Leipzig **84** (1932), 15–36.
4. Grötzsch, H., *Die Werte des Doppelverhältnisses bei schlichter konformer Abbildung*, Sitzungsber. Preuss. Akad. Wiss. Berlin (1933), 501–515.
5. Gunning, R.C., "Lectures on Riemann surfaces — Jacobi varieties," Princeton Univ. Press – Univ. of Tokyo Press, Princeton, 1972, 189pp.
6. Heins, M., *A problem concerning the continuation of Riemann surfaces*, in "Contributions to the theory of Riemann surfaces", ed. by L.V. Ahlfors et al. Princeton Univ. Press, Princeton, 1953, pp. 55–62.
7. Hurwitz, A. und Courant, R., "Vorlesungen über allgemeine Funktionentheorie und elliptische Funktionen. Geometrische Funktionentheorie," Virte Aufl. (mit einem Anhang von H. Röhrl), Berlin–Göttingen–Heidelberg–New York, 1964, 706pp.
8. Ioffe, M.S., *Some problems in the calculus of variations "in the large" for conformal and quasiconformal imbeddings of Riemann surfaces*, Sib. Math. J. **19** (1978), 587–603.
9. Jenkins, J.A., "Univalent functions and conformal mapping," Springer, Berlin–Göttingen–Heidelberg, 1958, 169pp.
10. Köditz, H. und Timmann, S., *Randschlichte meromorphe Funktionen auf endlichen Riemannschen Flächen*, Math. Ann. **217** (1975), 157–159.
11. Mori, A., *A remark on the prolongation of an open Riemann surface of finite genus*, J. Math. Soc. Japan **4** (1952), 27–30.
12. Oikawa, K., *On the prolongation of an open Riemann surface of finite genus*, Kodai Math. Sem Rep. **9** (1957), 34–41.
13. Oikawa, K., *On the uniqueness of the prolongation of an open Riemann surface of finite genus*, Proc. Amer. Math. Soc. **11** (1960), 785–787.
14. Rauch, H., *A transcendental view of the space of algebraic Riemann surfaces*, Bull. Amer. Math. Soc. **71** (1965), 1–39.
15. Renggli, H., *Structural instability and extensions of Riemann surfaces*, Duke Math. J. **42** (1975), 211–224.
16. Rodin, B. and Sario, L., "Principal functions," (with an appendix by M. Nakai), Van Nostrand, Princeton, 1968, 347pp.
17. Sario, L. and Oikawa, K., "Capacity functions," Springer, Berlin–Heidelberg–New York, 1969, 361pp.
18. Schiffer, M., *The span of multiply connected domains*, Duke Math. J. **10** (1943), 209–216.
19. Shiba, M., *The Riemann–Hurwitz relation, parallel slit covering map, and continuation of an open Riemann surface of finite genus*, Hiroshima Math. J. **14** (1984), 371–399.
20. Shiba, M., *The moduli of compact continuations of an open Riemann surface of genus one*, Trans. Amer. Math. Soc. **301** (1987), 299–311.
21. Shiba, M. and Shibata, K., *Hydrodynamic continuations of an open Riemann surface of finite genus*, Complex Variables Theory Appl. **8** (1987), 205–211.
22. Siegel, C.L., "Topics in complex function theory. Vol. II," Wiley–Interscience, New York, 1971, 193pp.
23. Springer, G., "Introduction to Riemann surfaces," Addison–Wesley, Reading, 1957, 309pp. (Reprint: Chelsea, New York, 1981).
24. Timmann, S., *Einbettungen endlicher Riemannscher Flächen*, Math. Ann. **217** (1975), 81–85.
25. Weil, A., *Modules des surfaces de Riemann*, Sem. Bourbaki, Exposé **168** (1957/58), 7pp.